Contents

How to use this guide

This guide is divided into two parts:

- **Part 1** Explains what fire risk assessment is and how you might go about it. Fire risk assessment should be the foundation for all the fire precautions in your premises.

- **Part 2** Provides further guidance on fire precautions. The information is provided for you and others to dip into during your fire risk assessment or when you are reviewing your precautions.

The appendices provide example checklists, some detailed technical information on fire-resisting elements and advice on historic buildings.

This guide is one from a series of guides listed on the back cover.

The rest of this introduction explains how the law applies.

Technical terms are explained in the glossary and references to other publications listed at the end of the publication are identified by a superscript number in the text.

In this guide reference is made to British Standards and standards provided by other bodies. The standards referred to are intended for guidance only and other standards could be used. Reference to any particular standard is not intended to confer a presumption of conformity with the requirements of the Regulatory Reform (Fire Safety) Order 2005 (the Order).[1]

The level of necessary safety (or service) must be dictated by the findings of your risk assessment so you may need to do more or less than that specified in any particular standard referred to. You must be prepared to show that what you have done complies with any requirements or prohibitions of the Order[1] irrespective of whether you have relied on a particular standard.

Preface

This guidance gives advice on how to avoid fires and how to ensure people's safety if a fire does start. Why should you read it? Because:

- Fire kills. In 2004 (England and Wales) fire and rescue services in England and Wales attended over 33,400 fires in non-domestic buildings. These fires killed 38 people and injured over 1,300.

- Fire costs money. The costs of a serious fire can be high and afterwards many businesses do not reopen. In 2004, the costs as a consequence of fire, including property damage, human casualties and lost business, were estimated at £2.5 billion.

This guide applies to England and Wales only. It does not set prescriptive standards, but provides recommendations and guidance for use when assessing the adequacy of fire precautions in small and medium places of assembly. Other fire risk assessment methods may be equally valid to comply with fire safety law. The guide also provides recommendations for the fire safety management of the premises.

Your existing fire safety arrangements may not be the same as the recommendations used in this guide but, as long as you can demonstrate that they meet an equivalent standard of fire safety, they are likely to be acceptable. If you decide that your existing arrangements are not satisfactory there may be other ways to comply with fire safety law. This means there is no obligation to adopt any particular solution in this guide if you prefer to meet the relevant requirement in some other way.

Where the building has been recently constructed or significantly altered, the fire detection and warning arrangements, escape routes and facilities for the fire and rescue service should have been designed, constructed and installed in accordance with current building regulations. In such cases, it is likely that these measures will be satisfactory as long as they are being properly maintained and no significant increase in risk has been introduced.

This guide should not be used to design fire safety in new buildings. Where alterations are proposed to existing premises, they may be subject to building regulations. However, it can be used to develop the fire safety strategy for the building.

Introduction

WHO SHOULD USE THIS GUIDE?

This guide is for all employers, managers, occupiers and owners of small and medium places of assembly. Details of other guides in the series are listed on the back cover. It tells you what you have to do to comply with fire safety law, helps you to carry out a fire risk assessment and identify the general fire precautions you need to have in place.

This guide is intended for premises where the main use of the building or part of the building is as a small (i.e. premises accommodating up to 60 people) or a medium (i.e. premises accommodating up to 300 people) place of assembly. These include: public houses; clubs; dance halls/schools; village halls; community centres; churches; other places of religious worship or study and associated premises; temporary structures and marquees/tents.

This guide does not apply to sports grounds or common areas of shopping malls; reference should be made to the guide for Large Places of Assembly for these premises.

This guide may also be suitable where the premises adjoin other complexes, although co-operation with the other managers will be necessary as part of an integrated risk assessment for the complex.

Also, where you handle and store flammable materials and substances, it will help you to take account of these in your risk assessment and help you to determine the necessary precautions to take to minimise the likelihood of them being involved in a fire.

The guide has been written to provide guidance for a responsible person, to help them to carry out a fire risk assessment in less complex premises. If you read the guide and decide that you are unable to apply the guidance, then you should seek expert advice from a competent person. More complex premises will probably need to be assessed by a person who has comprehensive training or experience in fire risk assessment. However, this guide can be used for multi-occupied buildings to address fire safety issues within their individual occupancies.

It may also be useful for:

- employees;

- employee-elected representatives;

- trade union-appointed health and safety representatives;

- enforcing authorities; and

- all other people who have a role in ensuring fire safety in small and medium places of assembly.

If your premises are listed as of historic interest, also see Appendix C.

Fire safety is only one of many safety issues management must address to minimise the risk of injury or death to staff or the public. Unlike most of the other safety concerns, fire has the potential to injure or kill large numbers of people very quickly. This guidance is concerned only with fire safety, but many of the measures discussed here will impact upon other safety issues, and vice versa. It is recognised that these differing safety demands can sometimes affect one another and management should consult other interested agencies, such as the Local Authority, where necessary to confirm that they are not contravening other legislation or guidance.

You can get advice about minimising fire losses from your insurer.

THE FIRE SAFETY ORDER

Previous general fire safety legislation

The Order[1] replaces previous fire safety legislation. Any fire certificate issued under the Fire Precautions Act 1971[2] will cease to have any effect. If a fire certificate has been issued in respect of your premises or the premises were built to recent building regulations, as long as you have made no material alterations and all the physical fire precautions have been properly maintained, then it is unlikely you will need to make any significant improvements to your existing physical fire protection arrangements to comply with the Order.[1] However, you must still carry out a fire risk assessment and keep it up to date to ensure that all the fire precautions in your premises remain current and adequate.

If you have previously carried out a fire risk assessment under the Fire Precautions (Workplace) Regulations 1997,[3] as amended 1999,[4] and this assessment has been regularly reviewed, then all you will need to do is revise that assessment taking account of the wider scope of the Order[1] as described in this guide.

Your premises may also be subject to the provisions of a licence or registration (e.g. under the Licensing Act 2003[47]), and the fire authority may wish to review your risk assessment as part of the licensing approval process. Fire safety conditions within your licence should not be set by a licensing authority where the Order[1] applies.

Background

The Order[1] applies in England and Wales. It covers general fire precautions and other fire safety duties which are needed to protect 'relevant persons' in case of fire in and around most 'premises'. The Order[1] requires fire precautions to be put in place 'where necessary' and to the extent that it is reasonable and practicable in the circumstances to do so.

Responsibility for complying with the Order[1] rests with the 'responsible person'. In a workplace, this is the employer and any other person who may have control of any part of the premises, e.g. the occupier or owner. In all other premises the person or people in control of the premises will be responsible. If there is more than one responsible person in any type of premises (e.g. a multi-occupied complex), all must take all reasonable steps to co-operate and co-ordinate with each other.

If you are the responsible person you must carry out a fire risk assessment which must focus on the safety in case of fire of all 'relevant persons'. It should pay particular attention to those at special risk, such as disabled people, those who you know have special needs and children, and must include consideration of any dangerous substance liable to be on the premises. Your fire risk assessment will help you identify risks that can be removed or reduced and decide the nature and extent of the general fire precautions you need to take.

If your organisation employs five or more people, your premises are licensed or an alterations notice is in force, you must record the significant findings of the assessment. It is good practice to record your significant findings in any case.

There are some other fire safety duties you need to comply with:

- **You must** appoint one or more competent persons, depending on the size and use of your premises, to carry out any of the preventive and protective measures required by the Order[1] (you can nominate yourself for this purpose). A competent person is someone with enough training and experience or knowledge and other qualities to be able to implement these measures properly.

- **You must** provide your employees with clear and relevant information on the risks to them identified by the fire risk assessment, about the measures you have taken to prevent fires, and how these measures will protect them if a fire breaks out.

- **You must** consult your employees (or their elected representatives) about nominating people to carry out particular roles in connection with fire safety and about proposals for improving the fire precautions.

- **You must**, before you employ a child, provide a parent with clear and relevant information on the risks to that child identified by the risk assessment, the measures you have put in place to prevent/protect them from fire and inform any other responsible person of any risks to that child arising from their undertaking.

- **You must** inform non-employees, such as temporary or contract workers, of the relevant risks to them, and provide them with information about who are the nominated competent persons, and about the fire safety procedures for the premises.

- **You must** co-operate and co-ordinate with other responsible persons who also have premises in the building, inform them of any significant risks you find, and how you will seek to reduce/control those risks which might affect the safety of their employees.

- **You must** provide the employer of any person from an outside organisation who is working in your premises (e.g. agency providing temporary staff) with clear and relevant information on the risks to those employees and the preventive and protective measures taken. You must also provide those employees with appropriate instructions and relevant information about the risks to them.

- If you are not the employer but have any control of premises which contain more than one workplace, **you are also responsible** for ensuring that the requirements of the Order[1] are complied with in those parts over which you have control.

- **You must** consider the presence of any dangerous substances and the risk this presents to relevant persons from fire.

- **You must** establish a suitable means of contacting the emergency services and provide them with any relevant information about dangerous substances.

- **You must** provide appropriate information, instruction and training to your employees, during their normal working hours, about the fire precautions in your workplace, when they start working for you, and from time to time throughout the period they work for you.

- **You must** ensure that the premises and any equipment provided in connection with firefighting, fire detection and warning, or emergency routes and exits are covered by a suitable system of maintenance, and are maintained by a competent person in an efficient state, in efficient working order and in good repair.

- **Your employees must** co-operate with you to ensure the workplace is safe from fire and its effects, and must not do anything that will place themselves or other people at risk.

The above outlines some of the main requirements of the Order.[1] The rest of this guide will explain how you might meet these requirements.

Responsibilities for short-term hiring or leasing and for shared use

Some premises or structures may be leased as an empty and unsupervised facility (e.g. village halls, temporary structures and marquees). The fire safety responsibilities of those leasing the building or structure (and, therefore, in charge of the activities conducted within the building or structure), and those of the owner/leasee, need to be established as part of the contract of hire.

In some premises (e.g. pubs), part of the premises may be hired out to another organisation for a separate function (e.g. DJs and dance promoters, weddings, birthdays and funeral organisers). The fire safety responsibilities of those organising the separate function, and those of the remainder of the building, need to be established as part of the contract of hire.

The responsible person for each individual unique, occasional or separate event or function will need to be clearly established and documented, and their legal duties made clear to them. In particular, and where necessary, the responsible person will need to take account of their own lack of familiarity with the layout of the premises, the fire safety provisions, and the duties of other responsible persons within the premises.

Who enforces the Fire Safety Order?

The local fire and rescue authority (the fire and rescue service) will enforce the Order[1] in most premises. The exceptions are:

- Crown-occupied/owned premises where Crown fire inspectors will enforce;

- premises within armed forces establishments where the defence fire and rescue service will enforce;

- certain specialist premises including construction sites, ships (under repair or construction) and nuclear installations, where the HSE will enforce; and

- sports grounds and stands designated as needing a safety certificate by the local authority, where the local authority will enforce.

The enforcing authority will have the power to inspect your premises to check that you are complying with your duties under the Order.[1] They will look for evidence that you have carried out a suitable fire risk assessment and acted upon the significant findings of that assessment. If you are required to record the outcome of the assessment they will expect to see a copy.

If the enforcing authority is dissatisfied with the outcome of your fire risk assessment or the action you have taken, they may issue an enforcement notice that requires you to make certain improvements or, in extreme cases, a prohibition notice that restricts the use of all or part of your premises until improvements are made.

If you are making changes to your premises, you should also consult your local licensing authority.

If your premises are considered by the enforcing authority to be or have potential to be high risk, they may issue an alterations notice that requires you to inform them before you make any changes to your premises or the way they are used.

Failure to comply with any duty imposed by the Order[1] or any notice issued by the enforcing authority is an offence. You have a right of appeal to a magistrates court against any notice issued. Where you agree that there is a need for improvements to your fire precautions but disagree with the enforcing authority on the technical solution to be used (e.g. what type of fire alarm system is needed) you may agree to refer this for an independent determination.

If having read this guide you are in any doubt about how fire safety law applies to you, contact the fire safety office at your local fire and rescue service.

If your premises were in use before 2006, then they may have been subject to the Fire Precautions Act[2] and the Fire Precautions (Workplace) Regulations.[3,4] Where the layout (means of escape) and other fire precautions have been assessed by the fire and rescue service to satisfy the guidance that was then current, it is likely that your premises already conform to many of the recommendations here, providing you have undertaken a fire risk assessment as required by the Fire Precautions (Workplace) Regulations.[3,4]

New buildings or significant building alterations should be designed to satisfy current building regulations[24] which address fire precautions. However, you will still need to carry out a fire risk assessment, or review your existing one (and act on your findings), to comply with the Order.[1]

Part 1 Fire risk assessment

MANAGING FIRE SAFETY

Good management of fire safety is essential to ensure that fires are unlikely to occur; that if they do occur they are likely to be controlled or contained quickly, effectively and safely; or that, if a fire does occur and grow, everyone in your premises is able to escape to a place of total safety easily and quickly.

The risk assessment that you must carry out will help you ensure that your fire safety procedures, fire prevention measures, and fire precautions (plans, systems and equipment) are all in place and working properly, and the risk assessment should identify any issues that need attention. Further information on managing fire safety is available in Part 2.

WHAT IS A FIRE RISK ASSESSMENT?

A fire risk assessment is an organised and methodical look at your premises, the activities carried on there and the likelihood that a fire could start and cause harm to those in and around the premises.

The aims of the fire risk assessment are:

• To identify the fire hazards.

• To reduce the risk of those hazards causing harm to as low as reasonably practicable.

• To decide what physical fire precautions and management arrangements are necessary to ensure the safety of people in your premises if a fire does start.

The term 'where necessary' (see Glossary) is used in the Order,[1] therefore when deciding what fire precautions and management arrangements are necessary you will need to take account of this definition.

The terms 'hazard' and 'risk' are used throughout this guide and it is important that you have a clear understanding of how these should be used.

• **Hazard:** anything that has the potential to cause harm.

• **Risk:** the chance of that harm occurring.

If your organisation employs five or more people, or your premises are licensed or an alterations notice requiring it is in force, then the significant findings of the fire risk assessment, the actions to be taken as a result of the assessment and details of anyone especially at risk must be recorded. You will probably find it helpful to keep a record of the significant findings of your fire risk assessment even if you are not required to do so.

HOW DO YOU CARRY OUT A FIRE RISK ASSESSMENT?

A fire risk assessment will help you determine the chances of a fire starting and the dangers from fire that your premises present for the people who use them and any person in the immediate vicinity. The assessment method suggested in this guide shares the same approach as that used in general health and safety legislation and can be carried out either as part of a more general risk assessment or as a separate exercise. As you move through the steps there are checklists to help you.

Before you start your fire risk assessment, take time to prepare, and read through the rest of Part 1 of this guide.

Much of the information for your fire risk assessment will come from the knowledge your employees, colleagues and representatives have of the premises, as well as information given to you by people who have responsibility for other parts of the building. A tour of your premises will probably be needed to confirm, amend or add detail to your initial views.

It is important that you carry out your fire risk assessment in a practical and systematic way and that you allocate enough time to do a proper job. It must take the whole of your premises into account, including outdoor locations and any rooms and areas that are rarely used. If your premises are small you may be able to assess them as a whole. In some premises you may find it helpful to divide them into a series of assessment areas using natural boundaries, e.g. assembly spaces, process areas (such as cooking facilities in village halls), offices, stores, as well as corridors, stairways and external routes.

If your premises are in a multi-use complex then the information on hazard and risk reduction will still be applicable to you. However, any alterations to the use or structure of your individual unit will need to take account of the overall fire safety arrangements in the building.

Your premises may be simple, with few people present or with a limited degree of activity, but if it forms part of a building with different occupancies, then the measures provided by other occupiers may have a direct effect on the adequacy of the fire safety measures in your premises.

Under health and safety law (enforced by the HSE or the local authority) you are required to carry out a risk assessment in respect of any work processes in your workplace and to take or observe appropriate special, technical or organisational measures. If your health and safety risk assessment identifies that these processes are likely to involve the risk of fire or the spread of fire then you will need to take this into account during your fire risk assessment under the Order,[1] and prioritise actions based on the level of risk.

You need to appoint one or more competent persons (this could be you) to carry out any of the preventive and protective measures needed to comply with the Order.[1] This person could be you, or an appropriately trained employee or, where appropriate, a third party.

Your fire risk assessment should demonstrate that, as far as is reasonable, you have considered the needs of all relevant people, including disabled people.

Figure 1 shows the five steps you need to take to carry out a fire risk assessment.

FIRE SAFETY RISK ASSESSMENT

1 Identify fire hazards

Identify:
Sources of ignition
Sources of fuel
Sources of oxygen

2 Identify people at risk

Identify:
People in and around the premises
People especially at risk

3 Evaluate, remove, reduce and protect from risk

Evaluate the risk of a fire occurring
Evaluate the risk to people from fire
Remove or reduce fire hazards
Remove or reduce the risks to people

- Detection and warning
- Fire-fighting
- Escape routes
- Lighting
- Signs and notices
- Maintenance

4 Record, plan, inform, instruct and train

Record significant finding and action taken
Prepare an emergency plan
Inform and instruct relevant people; co-operate and co-ordinate with others
Provide training

5 Review

Keep assessment under review
Revise where necessary

Remember to keep to your fire risk assessment under review.

Figure 1: The five steps of a fire risk assessment

STEP 1   IDENTIFY FIRE HAZARDS

For a fire to start, three things are needed:

• a source of ignition;

• fuel; and

• oxygen.

If any one of these is missing, a fire cannot start. Taking measures to avoid the three coming together will therefore reduce the chances of a fire occurring.

The remainder of this step will advise on how to identify potential ignition sources, the materials that might fuel a fire and the oxygen supplies that will help it burn.

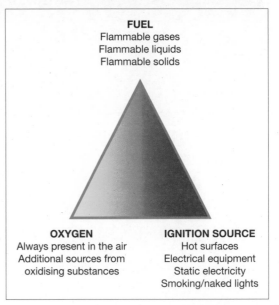

Figure 2: The fire triangle

1.1   Identify sources of ignition

You can identify the potential ignition sources in your premises by looking for possible sources of heat, which could get hot enough to ignite material found in your premises. These sources could include:

• cooking equipment, hot ducting, flues and filters;

• smokers' material, e.g. cigarettes, matches and lighters;

• electrical, gas or oil-fired heaters (fixed or portable), room heaters;

• hot processes, e.g. welding by contractors;

• faulty or misused electrical equipment;

• light fittings and lighting equipment, e.g. halogen lamps or display lighting;

• hot surfaces and obstruction of equipment ventilation;

• central heating boilers;

• naked flames, e.g. candles or gas or liquid-fuelled open-flame equipment;

• flares, fireworks and pyrotechnics; and

• arson.

Figure 3: Sources of ignition

Indications of 'near-misses', such as scorch marks on furniture or fittings, discoloured or charred electrical plugs and sockets, cigarette burns, etc., can help you identify hazards which you may not otherwise notice.

1.2   Identify sources of fuel

Anything that burns is fuel for a fire. You need to look for the things that will burn reasonably easily and are in enough quantity to provide fuel for a fire or cause it to spread to another fuel source. Some of the most common 'fuels' found in small and medium places of assembly are:

- flammable liquid-based products, such as paints, varnishes, thinners and adhesives;

- flammable liquids and solvents, such as alcohol (spirits), white spirit, methylated spirit, cooking oils and disposable cigarette lighters;

- flammable chemicals, such as certain cleaning products and photocopier chemicals;

- flammable gases such as liquefied petroleum gas (LPG) and acetylene;

- displays and stands;

- costumes, drapes and hangings, scenery and banners;

- packaged foodstuffs;

- packaging materials, stationery, advertising material and decorations;

- plastics and rubber, such as video tapes, polyurethane foam-filled furniture and polystyrene-based display materials and rubber or foam exercise mats;

- upholstered seating and cushions, textiles and soft furnishings and clothing displays;

- litter and waste products, particularly finely divided items such as shredded paper and wood shavings, offcuts, and dust accumulation amongst lubricated areas; and

- fireworks and pyrotechnics.

You should also consider the materials used to line walls and ceilings, e.g. polystyrene or carpet tiles, the fixtures and fittings, and brought-in materials, and how they might contribute to the spread of fire. Further information is available in Part 2, Section 1.

1.3   Identify sources of oxygen

The main source of oxygen for a fire is in the air around us. In an enclosed building this is provided by the ventilation system in use. This generally falls into one of two categories: natural airflow through doors, windows and other openings; or mechanical air conditioning systems and air handling systems. In many buildings there will be a combination of systems, which will be capable of introducing/extracting air to and from the building.

Additional sources of oxygen can sometimes be found in materials used or stored at premises such as:

• some chemicals (oxidising materials), which can provide a fire with additional oxygen and so help it burn. These chemicals should be identified on their container (and Control of Substances Hazardous to Health data sheet, see Figure 4) by the manufacturer or supplier who can advise as to their safe use and storage;

• oxygen supplies from cylinder storage; and

• pyrotechnics (fireworks) which contain oxidising materials and need to be treated with great care.

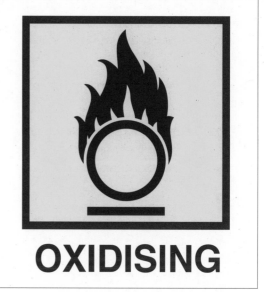

OXIDISING

Figure 4: Label on oxidising materials

Checklist

• Have you identified all potential ignition sources? ☐

• Have you identified all potential fuel sources? ☐

• Have you identified all potential sources of oxygen? ☐

• Have you made a note of your findings? ☐

STEP 2 IDENTIFY PEOPLE AT RISK

As part of your fire risk assessment, you need to identify those at risk if there is a fire. To do this you need to identify where you have people working, either at permanent locations (such as workstations) or at occasional locations around the premises, and to consider who else may be at risk, such as worshippers, customers, other users of the facilities, including visiting performance companies, visiting contractors, etc., and where these people are likely to be found.

You must consider all the people who use the premises, but you should pay particular attention to people who may be especially at risk such as:

• yourself and your helpers if you are 'managing' the premises but are unfamiliar with the location, e.g. organisers hiring the venue;

• employees who work alone and/or in isolated areas, e.g. cleaners and security staff;

• unaccompanied children;

• people who are unfamiliar with the premises, e.g. visitors and customers;

• people with disabilities* (including mobility impairment, or hearing or vision impairment, etc.);

*Visit the Disability Rights commission website on www.drc-gb.org for more information.

- people who may have some reason for not being able to leave the premises quickly, e.g. people in a state of undress (e.g. swimmers or performers), elderly customers, pregnant women or parents with children;

- sensory impaired due to alcohol, drugs or medication; and

- other people in the immediate vicinity of the premises.

In evaluating the risk to people with disabilities you may need to discuss their individual needs with them. In more complex buildings used extensively by the public you may need to seek professional advice.

Further guidance on help for people with special needs is given in Part 2, Section 1.13.

Checklist

- Have you identified who is at risk?   ☐

- Have you identified why they are at risk?   ☐

- Have you made a note of your findings?   ☐

STEP 3   EVALUATE, REMOVE, REDUCE AND PROTECT FROM RISK

The management of the premises and the way people use it will have an effect on your evaluation of risk. Management may be your responsibility alone or there may be others, such as the building owners or managing agents, who also have responsibilities. In multi-occupied buildings all those with some control must co-operate and you need to consider the risk generated by others in the building.

3.1   Evaluate the risk of a fire occuring

The chances of a fire starting will be low if your premises has few ignition sources and combustible materials are kept away from them.

In general, fires start in one of three ways:

- accidentally, such as when smoking materials are not properly extinguished or when mobile heaters are knocked over;

- by act or omission, such as when electrical equipment is not properly maintained, or when waste is allowed to accumulate near to a heat source; and

- deliberately, such as an arson attack involving setting fire to external rubbish bins placed too close to the building.

Look critically at your premises and try to identify any accidents waiting to happen and any acts or omissions which might allow a fire to start. You should also look for any situation that may present an opportunity for an arsonist.

Further guidance is given in Part 2, Section 1 on evaluating the risk of a fire starting.

3.2   Evaluate the risk to people

In Step 2 you identified the people likely to be at risk should a fire start anywhere in the premises and earlier in Step 3 you identified the chances of a fire occurring. It is unlikely that you will have concluded that there is no chance of a fire starting anywhere in your premises so you now need to evaluate the actual risk to those people should a fire start and spread from the various locations that you have identified.

While determining the possible incidents, you should also consider the likelihood of any particular incident; but be aware that some very unlikely incidents can put many people at risk.

To evaluate the risk to people in your premises, you will need to understand the way fire can spread. Fire is spread by three methods:

• convection;

• conduction; and

• radiation.

Convection

Fire spread by convection is the most dangerous and causes the largest number of injuries and deaths. When fires start in enclosed spaces such as buildings, the smoke rising from the fire gets trapped by the ceiling and then spreads in all directions to form an ever-deepening layer over the entire room space. The smoke will pass through any holes or gaps in the walls, ceiling and floor into other parts of the building. The heat from the fire gets trapped in the building and the temperature rises.

Conduction

Some materials, such as metal shutters and ducting, can absorb heat and transmit it to the next room, where it can set fire to combustible items that are in contact with the heated material.

Radiation

Radiation heats the air in the same way as an electric bar heater heats a room. Any material close to a fire will absorb the heat until the item starts to smoulder and then burn.

Smoke produced by a fire also contains toxic gases which are harmful to people. A fire in a building with modern fittings and materials generates smoke that is thick and black, obscures vision, causes great difficulty in breathing and can block the escape routes.

Figure 5: Smoke moving through a building

It is essential that the means of escape and other fire precautions are adequate to ensure that everyone can make their escape to a place of total safety before the fire and its effects can trap them in the building.

In evaluating this risk to people you will need to consider situations such as:

- fire starting within an assembly space, such as a church hall, where many people could be affected;

- fire starting on a lower floor affecting the only escape route for people on upper floors or the only escape route for people with disabilities;

- fire developing in an unoccupied space that people have to pass by to escape from the building;

- fire or smoke spreading through a building via routes such as vertical shafts, service ducts, ventilation systems, poorly installed, poorly maintained or damaged walls, partitions and ceilings affecting people in remote areas;

- fire spreading rapidly through the building because of combustible structural elements and/or large quantities of combustible goods;

- fire and smoke spreading through a building due to poor installation of fire precautions, e.g. incorrectly installed fire doors (see Appendix B2 for more information on fire doors) or incorrectly installed services penetrating fire walls;

- fire and smoke spreading through the building due to poorly maintained and damaged fire doors or fire doors being wedged open; and

- fire starting in a service room and affecting hazardous materials (such as pyrotechnics or swimming pool chemicals).

Further guidance on fire risks is given in Part 2, Section 1.

3.3   Remove or reduce the hazards

Having identified the fire hazards in Step 1, you now need to remove those hazards if reasonably practicable to do so. If you cannot remove the hazards, you need to take reasonable steps to reduce them if you can. This is an essential part of fire risk assessment and as a priority this must take place before any other actions.

Ensure that any actions you take to remove or reduce fire hazards or risk are not substituted by other hazards or risks. For example, if you replace a flammable substance with a toxic or corrosive one, you must consider whether this might cause harm to people in other ways.

Remove or reduce sources of ignition

There are various ways that you can reduce the risk caused by potential sources of ignition, for example:

- Wherever possible replace a potential source by a safer alternative.

- Replace naked flame and radiant heaters with fixed convector heaters or a central heating system. Fire guard naked flames and restrict the movement of portable heating appliances.

- Restrict and control the use of naked flames, e.g. candles.

- Operate a safe smoking policy in designated smoking areas, ensuring sufficient ashtrays are provided and cleaned appropriately, and prohibit smoking elsewhere.

- Ensure that sources of heat are kept away from flammable materials such as curtains, scenery and displays.

- Ensure electrical, mechanical and gas equipment is installed, used, maintained and protected in accordance with the manufacturer's instructions.

- Strictly control hot proceses/hot work by operating permit to work schemes.

- Ensure that no one carrying out work on gas fittings which involves exposing pipes that contain or have contained flammable gas uses any source of ignition such as blow-lamps or hot-air guns.

- Ensure that no-one uses any source of ignition while searching for an escape of gas.

- Ensure cooking and catering equipment is installed, used, maintained and protected in accordance with the manufacturer's instructions.

- Check all areas where hot work (e.g. welding) has been carried out, to ensure that no ignition has taken place or any smouldering materials remain that may cause a fire.

- Ensure that flares and fireworks are not brought into the premises by members of the public.

- Ensure that all pyrotechnics, fireworks, flares and other hazardous equipment are installed, used, maintained, protected and stored in accordance with the manufacturer's instructions.

- Take precautions to avoid arson.

Remove or reduce sources of fuel

There are various ways that you can reduce the risks caused by materials and substances which burn, for example:

- Ensure that all upholstered furniture, curtains, drapes and other soft furnishings, are fire-retardant, or have been treated with a proprietory fire-retardant treatment designed to enhance their fire performance.

- Ensure that display materials (including artificial and dried foliage), scenery and stands, are fire-retardant, or have been treated with a proprietory fire-retardant treatment designed to enhance their fire performance.

- Reduce stocks of flammable materials, liquids and gases on display in public areas to a minimum. Keep remaining stock in dedicated storerooms or storage areas, preferably outside, where the public are not allowed to go, and keep the minimum required for the operation of the business.

- Ensure flammable materials, liquids and gases, are kept to a minimum, and are stored properly with adequate separation distances between them.

- Do not keep flammable solids, liquids and gases together.

- Remove, cover or treat large areas of highly combustible wall and ceiling linings, e.g. polystyrene or carpet tiles, to reduce the rate of flame spread across the surface.

- Develop a formal system for the control of combustible waste by ensuring that waste materials and rubbish are not allowed to build up and are carefully stored until properly disposed of, particularly at the end of the day.

- Ensure that foam mats (e.g. gymnasium mats), contents of foam pits and similar equipment are of combustion modified foam. Cover pits when not in use.

- Do not keep scenery or properties which are not in current use on an open stage other than in an approved scenery store or property store.

- Ensure that quantities of all pyrotechnics, fireworks and other hazardous equipment are kept to a minimum.

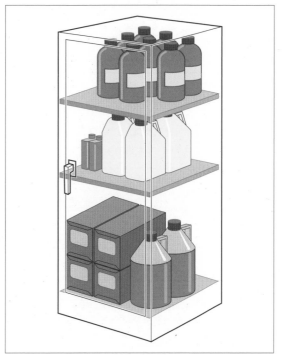

Figure 6: Storage of flammables

- Take action to avoid any parts of the premises, and in particular storage areas and ground staff equipment, being vulnerable to arson or vandalism.

- Check all areas where hot work (e.g. welding) has been carried out to ensure that no ignition has taken place and no smouldering or hot materials remain that may cause a fire later.

Further guidance on removing and reducing hazards is given in Part 2, Section 1.

Remove or reduce sources of oxygen

You can reduce the potential source of oxygen supplied to a fire by:

- closing all doors, windows and other openings not required for ventilation, particularly out of working hours;

- shutting down ventilation systems which are not essential to the function of the premises;

- not storing oxidising materials (including pyrotechnics and fireworks) near or with any heat source or flammable materials;

- controlling the use and storage of oxygen cylinders, ensuring that they are not leaking, are not used to 'sweeten' the atmosphere, and that where they are located is adequately ventilated; and

- controlling the use and storage of pyrotechnics and fireworks.

3.4 Remove or reduce the risks to people

Having evaluated and addressed the risk of fire occuring and the risk to people (preventative measures) it is unlikely that you will be able to conclude that no risk of fire starting and presenting a risk remains to people in your premises.

You now need to reduce any remaining fire risk to people to as low as reasonably practicable, by ensuring that adequate fire precautions are in place to warn people in the even of fire and allow them to escape safely.

The rest of this step describes the fire protection measures you may wish to adopt to reduce the remaining fire risk to people (see Steps 3.4.1 to 3.4.6).

The level of fire protection you need to provide will depend on the level of risk that remains in the premises after you have removed or reduced the hazards and risks. Part 2, Section 4.1 can help you decide the level of risk that you may still have.

Flexibility of fire protection measures

Flexibility will be required when applying this guidance; the level of fire protection should be proportional to the risk posed to the safety of the people in the premises. Therefore, the objective should be to reduce the remaining risk to a level as low as reasonably practicable. The higher the risk of fire and risk to life, the higher the standards of fire protection will need to be.

Your premises may not exactly fit the solutions suggested in this guide and they may need to be applied in a flexible manner without compromising the safety of the occupants.

For example, if the travel distance is in excess of the norm for the level of risk you have determined (see Part 2, Table 2 on page 65), it may be necessary to do any one or a combination of the following to compensate:

- Provide earlier warning of fire using automatic fire detection.

- Revise the layout to reduce travel distances.

- Reduce the fire risk by removing or reducing combustible materials and/or ignition sources.

- Control the number of people in the premises.

- Limit the area to trained staff only (no public).

- Increase staff training and awareness.

Note: The above list is not exhaustive and is only used to illustrate some examples of trade-offs to provide safe premises.

If you decide to significantly vary away from the benchmarks in this guidance then you should seek expert advice before doing so.

3.4.1 Fire detection and warning systems

In many premises a fire may be obvious to everyone as soon as it starts (e.g. in a simple open plan village hall). In these cases, where the number and position of exits and the travel distance to them is adequate, a simple shout of 'fire' or a simple manually operated device, such as a gong, whistle or air horn that can be heard by everybody when operated from any single point within the building, may be all that is needed.

Figure 7: Fire detection and warning system

Organisors or staff will need a managed fire evacuation plan to do this.

Where an alarm given from any single point is unlikely to be heard throughout the building, an electrical system incorporating sounders and manually operated call points (break-glass boxes) is likely to be required. This type of system is likely to be acceptable where all parts of the building are occupied at the same time and it is unlikely that a fire could start without somebody noticing it quickly. However, where there are unoccupied areas, or common corridors and circulation spaces in multi-occupied buildings, in which a fire could develop to the extent that escape routes could be affected before the fire is discovered, an automatic fire detection system with a control panel which is able to identify the zone or specific location where the alarm has been raised may be necessary (see Figure 7).

You may need to consider special arrangements for times when people are working alone, are disabled, or when your normal occupancy patterns are different, e.g. when maintenance staff or other contractors are working while the premises are closed.

In complex premises, particularly those accommodating large numbers of people in different rooms, such as community centres, it is likely that a more sophisticated form of warning and evacuation, possibly involving staff alarms, should be provided (where staff are alerted before members of the public).

The control panel (or a repeater) should be located near the entrance to the premises. Arrangements should be made to brief the fire and rescue service when they arrive.

False alarms from electrical fire warning systems are a major problem and result in many unwanted calls to the fire and rescue service each year. To help reduce the number of false alarms, the design and location of activation devices should be reviewed against the way the premises are currently used.

If you are not sure whether your current arrangements are adequate, see the additional guidance on fire warning systems in Part 2, Section 2.

Checklist

- Can the existing means of detection ensure a fire is discovered quickly enough for the alarm to be raised in time for all the occupants to escape to a place of total safety? ☐

- Are the detectors of the right type and in the appropriate locations ☐

- Can the means of warning be clearly heard and understood by everyone throughout the whole building when initiated from a single point? Are there provisions in locations where the alarm cannot be heard? ☐

- If the fire detection and warning system is electrically powered, does it have a back-up power supply? ☐

3.4.2   Firefighting equipment and facilities

Firefighting equipment can reduce the risk of a small fire, e.g. a fire in a waste-paper bin, developing into a large one. The safe use of an appropriate fire extinguisher to control a fire in its early stages can also significantly reduce the risk to other people in the premises by allowing people to assist others who are at risk.

This equipment will need to comprise enough portable extinguishers that must be suitable for the risk.

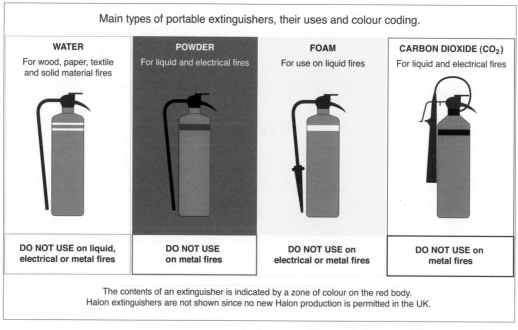

Main types of portable extinguishers, their uses and colour coding.

WATER	POWDER	FOAM	CARBON DIOXIDE (CO$_2$)
For wood, paper, textile and solid material fires	For liquid and electrical fires	For use on liquid fires	For liquid and electrical fires
DO NOT USE on liquid, electrical or metal fires	DO NOT USE on metal fires	DO NOT USE on electrical or metal fires	DO NOT USE on metal fires

The contents of an extinguisher is indicated by a zone of colour on the red body.
Halon extinguishers are not shown since no new Halon production is permitted in the UK.

Figure 8: Types of fire extinguishers

In simple premises, having one or two portable extinguishers of the appropriate type, readily available for use, may be all that is necessary.

In complex premises, such as multi-roomed community centres, where it is considered that a number of portable extinguishers are required, they should be sited in suitable locations, such as on the escape routes at each floor level. It may also be necessary to indicate the location of extinguishers with suitable signs. You should locate extinguishers in areas where they can be easily accessed by trained members of staff, but not in areas open to misuse or vandalism.

Some premises will also have permanently installed firefighting equipment such as hose reels, for use by trained staff or firefighters.

People with no training should not be expected to attempt to extinguish a fire. However, all staff should be familiar with the location and basic operating procedures for the equipment provided, in case they need to use it. If your fire strategy means that certain people, e.g. fire marshals, will be expected to take a more active role, then they should be provided with more comprehensive training.

Other fixed installations and facilities to assist firefighters, such as dry rising mains and access for fire engines, or automatically operated, fixed fire suppression systems such as sprinklers and gas or foam flooding systems, may also have been provided.

Where these have been required by law, e.g. the Building Regulations or local Acts, such equipment and facilities must be maintained.

Similarly, if provided for other reasons, e.g. insurance, it is good practice to ensure that they are properly maintained.

In most cases it will be necessary to consult a competent service engineer. Further information is given in Part 2, Section 3. Keeping records of the maintenance carried out will help you demonstrate to the enforcing authority that you have complied with fire safety law.

Appendix A.1 provides a sample fire safety maintenance checklist you can use.

For more guidance on portable extinguishers see Part 2, Section 3.1, for fixed firefighting installations, Part 2, Section 3.2 and other facilities (including those for firefighters) see Part 2, Section 3.3.

Checklist

- Are the extinguishers suitable for the purpose? ☐

- Are there enough extinguishers sited throughout the premises at appropriate locations? ☐

- Are the right types of extinguishers located close to the fire hazards and can users get to them without exposing themselves to risk? ☐

- Are the extinguishers visible or does their position need indicating? ☐

- Have you taken steps to prevent the misuse of extinguishers? ☐

- Do you regularly check equipment provided to help maintain the escape routes? ☐

- Do you carry out daily checks to ensure that there is clear access for fire engines? ☐

- Are those who test and maintain the equipment competent to do so? ☐

- Do you have the necessary procedures in place to maintain any facilities that have been provided for the safety of people in the building (or for the use of firefighters, such as access for fire engines and firefighting lifts)? ☐

3.4.3　Escape routes

Once a fire has started, been detected and a warning given, everyone in your premises should be able to escape to a place of total safety unaided and without the help of the fire and rescue service. However, some people with disabilities and others with special needs may need help from staff who will need to be designated for the purpose.

Escape routes should be designed to ensure, as far as possible, that any person confronted by fire anywhere in the building should be able to turn away from it and escape to a place of reasonable safety, e.g. a protected stairway. From there they will be able to go directly to a place of total safety away from the building.

Those who require special assistance (e.g. very young children and some people with disabilities) could be accommodated on the same level as the final exit from the premises to facilitate escape. Where they need assistance to evacuate, you should make sure that there are sufficient staff to ensure a speedy evacuation.

The level of fire protection that should be given to escape routes will vary depending on the level of risk of fire within the premises and other related factors.

When determining whether your premises have adequate escape routes, you need to consider a number of factors, including:

* the type and number of people using the premises;

* escape time;

* the age and construction of the premises;

* the number and complexity of escape routes and exits;

* assembly points;

* assisted means of escape/personal evacuation plans;

* whether lifts can or need to be used; and

* the use of delayed alarms (staff alarms).

The type and number of people using the premises

The people present in your premises will sometimes be employees (out of opening hours), but most of the time will be a mixture of employees and members of the public (including disabled people and unaccompanied children). Employees can reasonably be expected to have an understanding of the layout of the premises, while members of the public, particularly in more complex premises, will be unlikely to have knowledge of alternative escape routes.

The number and capability of people present will influence your assessment of the escape routes. You must ensure that your existing escape routes are sufficient and capable of safely evacuating all the people likely to use your premises at any time, and particularly during times of high usage (e.g. during an amateur dramatics perfomance in a village hall). If necessary you may need either to increase the capacity of the escape routes or restrict the number of people in the premises.

In considering the numbers of people using your premises, you may also need to take account of the possible effects of intoxicants (alcohol, illegal drugs, etc.) on how people may react in a fire. You also need to take account of the type of event being held on the premises, for example if it is a ticketed event, or an event with free admission. You may have to consider monitoring the number of people entering the premises, controlling entry numbers or controlling the numbers within specific areas, and/or preventing migration and overcrowding. You may also wish to consider other behavioural issues such as parents seeking their children and people reluctant to abandon their property.

Escape time

In the event of a fire, it is important to evacuate people as quickly as possible from the premises. Escape routes in a building should be designed so that people can escape quickly enough to ensure that they are not placed in any danger from fire. The time available will depend on a number of factors, including how quickly the fire is detected and the alarm raised, the number of escape routes available, the nature of the occupants and the speed of fire growth. For simplicity, the travel distances in Part 2, Table 2 on page 65 take these factors into account. Part 2, Section 4.1 will help you decide the level of risk in your premises for escape purposes.

The age and construction of the premises

Older buildings may comprise different construction materials from newer buildings and may be in a poorer state of repair. The materials from which your premises are constructed, the quality of building work and state of repair could contribute to the speed with which any fire may spread, and potentially affect the escape routes the occupants will need to use. A fire starting in a building constructed mainly from combustible material will spread faster than one where fire-resisting construction materials have been used.

If you wish to construct internal partitions or walls in your premises, perhaps to divide up a meeting room, you should ensure that any new partition or wall does not obstruct any escape routes or fire exits, extend travel distances or reduce the sound levels of the fire alarm system. Any walls that affect the means of escape should be constructed of appropriate material. (Further technical information is provided in Appendix B.)

Depending on the findings of your fire risk assessment, it may be necessary to protect the escape routes against fire and smoke by upgrading the construction of the floors, ceiling and walls to be a fire-resisting standard. You should avoid having combustible wall and ceiling linings in your escape routes. For further information see Appendix B. You may need to seek advice from a competent person. Any structural alterations may require building regulation approval.

The number of escape routes and exits

In general there should normally be at least two escape routes from all parts of the premises, but a single escape route may be acceptable in some circumstances (e.g. a premises or part of a premises accommodating less than 60 people or where the travel distances are limited).

Where two escape routes are necessary and to further minimise the risk of people becoming trapped, you should ensure that the escape routes are completely independent of each other. This will prevent a fire affecting more than one escape route at the same time.

When evaluating escape routes, you may need to build in a safety factor by discounting the largest exit from your escape plan or doors which cannot be opened quickly. You can then determine whether the remaining escape routes from a room, floor or building will be sufficient to evacuate all the occupants within a reasonable time. Escape routes provide that escape in a single direction only may need additional fire precautions to be regarded as adequate.

Exit doors on escape routes and final exit doors should normally open in the direction of travel, and be quickly and easily openable without the need for a key. Checks should be made to ensure final exits are wide enough to accommodate the number of people who may use the escape routes they serve.

Management of escape routes

It is essential that escape routes, and the means provided to ensure they are used safely, are managed and maintained to ensure that they remain usable and available at all times when the premises are occupied. Tell employees in staff training sessions about the escape routes within the premises.

Figure 9: A blocked corridor with incorrect signage

Corridors and stairways that form part of escape routes should be kept clear and hazard free at all times. Items that may be a source of fuel or pose an ignition risk should never be located on any corridor or stairway that will be used as an escape route.

Further guidance is available in Part 2, Section 4.

Emergency evacuation of persons with mobility impairment

The means of escape you provide must be suitable for the evacuation of everyone likely to be in your premises. This may require additional planning and allocation of staff roles – with appropriate training. Provisions for the emergency evacuation of disabled persons may include:

• stairways;

• evacuation lifts;

• firefighting lifts;

• horizontal evacuation;

• refuges; and

• ramps.

Use of these facilities will need to be linked to effective management arrangements as part of your emergency plan. The plan should not rely on fire and rescue service involvement for it to be effective.

Marquees, tents and temporary structures

Exit routes from marquees, tents and temporary structures may be over uneven ground or temporary flooring, duck boards, ramps, etc. These factors should be taken into account to ensure that there are safe egress routes. Travel distances should be shorter than in conventional buildings (see Part 2, Section 4.1).

Further guidance on escape routes is available in Part 2, Section 4.

Checklist

- Is your building constructed, particularly in the case of multi-storey buildings, so that, if there is a fire, heat and smoke will not spread uncontrolled through the building to the extent that people are unable to use the escape routes? ☐

- Are any holes or gaps in walls, ceilings and floors properly sealed, e.g. where services such as ventilation ducts and electrical cables pass through them? ☐

- Can all the occupants escape to a place of total safety in a reasonable time? ☐

- Are the existing escape routes adequate for the numbers and type of people that may need to use them e.g. staff and members of the public, including disabled people? ☐

- Are the exits in the right place and do the escape routes lead as directly as possible to a place of total safety? ☐

- If there is a fire, could all available exits be affected or will at least one route from any part of the premises remain available? ☐

- Are the escape routes and final exits kept clear at all times? ☐

- Do the doors on escape routes open in the direction of escape? ☐

- Can all final exit doors be opened easily and immediately if there is an emergency? ☐

- Will everybody be able to safely use the escape routes from your premises? ☐

- Are the people who work in the building aware of the importance of maintaining the safety of the escape routes, e.g. by ensuring that fire doors are not wedged open and that combustible materials are not stored within escape routes? ☐

- Are there any particular or unusual issues to consider? ☐

3.4.4 Emergency escape lighting

People in your premises must be able to find their way to a place of total safety if there is a fire by using escape routes that have enough lighting. Where any escape routes are internal and without windows, or your premises are used during periods of darkness, including early darkness on winter days, then some form of back-up to the normal escape route lighting (emergency escape lighting) is likely to be required.

In simple premises, where the escape routes are straightforward, borrowed lighting, e.g. from street lamps where they illuminate escape routes, may be acceptable. Where borrowed lighting is not available, suitably placed torches may be acceptable for the use of staff only.

In complex premises it is likely that a more comprehensive system of electrical automatic emergency escape lighting will be needed to illuminate all the escape routes.

Where people have difficulty seeing conventional signs, a 'way-guidance' system may need to be considered.

Further guidance on emergency escape lighting is given in Part 2, Section 5.

Checklist

- Are your premises used during periods of darkness? ☐
- Will there always be sufficient lighting to safely use escape routes? ☐
- Do you have back-up power supplies for your emergency lighting? ☐

3.4.5 Signs and notices

Signs

Signs must be used, where necessary, to help people identify escape routes, find firefighting equipment and emergency fire telephones. These signs are required under the Health and Safety (Safety Signs and Signals) Regulations 1996[5, 6] and must comply with the provisions of those Regulations.

A fire risk assessment that determines that no escape signs are required (because, for example, trained staff will always be available to help persons to escape routes), is unlikely to be acceptable to an enforcing authority other than in the smallest and simplest of premises where the exits are in regular use and familiar to staff and guests (e.g. in a small village hall).

For a sign to comply with these Regulations it must be in pictogram form (see Figure 10). The pictogram can be supplemented by text if this is considered necessary to make the sign more easily understood, but you must not have a safety sign that uses only text.

Where the locations of escape routes and firefighting equipment are readily apparent and the firefighting equipment is visible at all times, then signs are not necessary. In all other situations it is likely that the fire risk assessment will indicate that signs will be necessary.

Figure 10: Typical fire exit sign

Notices

Notices must be used, where necessary, to provide the following:

- instructions on how to use any fire safety equipment;

- the actions to be taken in the event of fire; and

- help for the fire and rescue service (e.g. location of sprinkler valves or electrical cut-off switches).

All signs and notices should be positioned so that they can be easily seen and understood.

Figure 11: Simple fire action notice

Further guidance on signs and notices is given in Part 2, Section 6.

Checklist

- Where necessary, are escape routes and exits, the locations of firefighting equipment and emergency telephones indicated by appropriate signs? ☐

- Have you provided notices such as those giving information on how to operate security devices on exit doors, those indicating doors enclosing fire hazards that must be kept shut and fire action notices for staff and other people? ☐

- Are you maintaining all the necessary signs and notices so that they continue to be correct, legible and understood? ☐

- Are you maintaining signs that you have provided for the information of the fire and rescue service, such as those indicating the location of water suppression stop valves and the storage of hazardous substances? ☐

3.4.6   Installation, testing and maintenance

New fire precautions should be installed by a competent person.

You must keep any existing equipment, devices or facilities that are provided in your premises for the safety of people, such as fire alarms, fire extinguishers, lighting, signs, fire exits and fire doors, in effective working order and maintain separating elements designed to prevent fire and smoke entering escape routes.

You must ensure regular checks, periodic servicing and maintenance are carried out whatever the size of your premises and any defects are put right as quickly as possible.

You, or a person you have nominated, can carry out certain checks and routine maintenance work. Further maintenance may need to be carried out by a competent service engineer. Where contractors are used, third party certification is one method where a reasonable assurance of quality of work and competence can be achieved (see Part 2, Section 8).

If you allow your premises to be hired (e.g. a village hall) you retain overall responsibility. However, some of the checking responsibilities should be passed to the hirer under their hiring agreement to carry out these checks on the day they use the premises.

The following are examples of checks and tests that should be considered. You should determine the appropriate period for these checks from your risk assessment. The examples of testing and maintenance given are not intended to be prescriptive and other regimes may be appropriate.

Daily checks (and/or checks before an event)

Remove bolts, padlocks and security devices from fire exits ensure that doors on escape routes swing freely and close fully, and check escape routes to ensure they are clear from obstructions and combustible materials, and in a good state of repair. Check the fire alarm panel to ensure the system is active and fully operational. Where practicable, visually check that emergency lighting units are in good repair and apparently working. Check that all safety signs and notices are legible. (See Appendix B3 for more details on bolts, padlocks and security devices.)

Weekly tests and checks

Test fire detection and warning systems and manually-operated warning devices weekly following the manufacturer's or installer's instructions. Check the batteries of safety torches and that fire extinguishers and hose reels are correctly located and in apparent working order.

Monthly tests and checks

Test all emergency lighting systems and safety torches to make sure they have enough charge and illumination according to the manufacturer's or supplier's instructions. This should be at an appropriate time when, following the test, they will not be immediately required.

Check that fire doors are in good working order and closing correctly and that the frames and seals are intact.

Six-monthly tests and checks

A competent person should test and maintain the fire-detection and warning system.

Annual tests and checks

The emergency lighting and all firefighting equipment, fire alarms and other installed systems should be tested and maintained by a competent person.

All structural fire protection and elements of fire compartmentation should be inspected and any remedial action carried out. Specific guidance on the maintenance of timber fire-resisting doors is given in Appendix B2.

Appendix A.1 provides an example of a fire safety maintenance checklist. You will find it useful to keep a log book of all maintenance and testing.

Further guidance on maintenance and testing of individual types of equipment and facilities can be found in the relevant section in Part 2.

Checklist

- Do you regularly check all fire doors and escape routes and associated lighting and signs? ☐
- Do you regularly check all your firefighting equipment? ☐
- Do you regularly check your fire detection and alarm equipment? ☐
- Are those who test and maintain the equipment competent to do so? ☐
- Do you keep a log book to record tests and maintenance? ☐

Step 3 Checklist

Evaluate, remove, reduce and protect from risks by:

- Evaluating the risk to people in your building if a fire starts. ☐
- Removing or reducing the hazards that might cause a fire. ☐

  Have you:

  – Removed or reduced sources of ignition? ☐

  – Removed or reduced sources of fuel? ☐

  – Removed or reduced sources of air or oxygen? ☐

  Have you removed or reduced the risks to people if a fire occurs by:

  – Considering the need for fire detection and for warning? ☐

  – Considering the need for firefighting equipment? ☐

  – Determining whether your escape routes are adequate? ☐

  – Determining whether your lighting and emergency lighting are adequate? ☐

  – Checking that you have adequate signs and notices? ☐

  – Regularly testing and maintaining safety equipment? ☐

  – Considering whether you need any other equipment or facilities? ☐

STEP 4   RECORD, PLAN, INFORM, INSTRUCT AND TRAIN

In Step 4 there are four further elements of the risk assessment you should focus on to address the management of fire safety in your premises. In some premises with simple layouts this could be done as part of the day-to-day management; however, as the premises or the organisation get larger it may be necessary for a formal structure and written policy to be developed. Further guidance on managing fire safety is given in Part 2 on page 39.

4.1   Record the significant findings and action taken

If you or your organisation employ five or more people, your premises are licensed, or an alterations notice requiring you to do so is in force, you must record the significant findings of your fire risk assessment and the actions you have taken.

Significant findings should include details of:

- the fire hazards you have identified (you don't need to include trivial things like a small tin of solvent-based glue);

- the actions you have taken or will take to remove or reduce the chance of a fire occurring (preventive measures);

- persons who may be at risk, particularly those especially at risk;

- the actions you have taken or will take to reduce the risk to people from the spread of fire and smoke (protective measures);

- the actions people need to take in case of fire including details of any persons nominated to carry out a particular function (your emergency plan); and

- the information, instruction and training you have identified that people need and how it will be given.

For further information, see Part 2.

You may also wish to record discussions you have had with staff or staff representatives (including trade unions).

Even where you are not required to record the significant findings, it is good practice to do so.

In some premises record keeping may be no more than a few sheets of paper (possibly forming part of a health and safety folder), containing details of significant findings, any action taken and a copy of the emergency plan.

The record could take the form of a simple list which may be supported by a simple plan of the premises (see Figure 12). More complex premises may need a more detailed record; however, there is no one 'correct' format specified for this.

You must be able to satisfy the enforcing authority, if called upon to do so, that you have carried out a suitable and sufficient fire risk assessment. Keeping records will help you do this and will also form the basis of your subsequent reviews. If you keep records, you do not need to record all the details, only those that are significant and the action you have taken.

It might be helpful to include a simple line drawing. This can also help you check your fire precautions as part of your ongoing review.

Figure 12: Example of a line drawing showing general fire safety precautions

The findings of your fire risk assessment will help you to develop your emergency plan, the instruction, information and training you need to provide, the co-operation and co-ordination arrangements you may need to have with other responsible people and the arrangements for maintenance and testing of the fire precautions. If you are required to record the significant findings of your fire risk assessment then these arrangements must also be recorded.

Further guidance about fire safety records with an example is given in Part 2, Section 7.1.

Checklist

- Have you recorded the significant findings of your assessment? ☐
- Have you recorded what you have done to remove or reduce the risk? ☐
- Are your records available for inspection by the enforcing authority? ☐

4.2 Emergency plans

You need to have an emergency plan for dealing with an fire situation.

The purpose of an emergency plan is to ensure that the people in your premises know what to do if there is a fire and that the premises can be safely evacuated.

If you or your organisation employ five or more people, or your premises are licensed or an alterations notice requiring it is in force, then details of your emergency plan must be recorded. Even if it is not required, it is good practice to keep a record.

Your emergency plan should be based on the outcome of your fire risk assessment and be available for your employees, their representatives (where appointed) and the enforcing authority.

In premises with simple layouts the emergency plan may be no more than a fire action notice. In multi-occupied and more complex premises the emergency plan will need to be more detailed and compiled after consultation with other occupiers and other responsible people, e.g. owners, who have control over the building. In most cases this means that an emergency plan covering the whole building will be necessary. It will help if you can agree one person to co-ordinate this task.

Further guidance on emergency plans is given in Part 2, Section 7.2.

Checklist

- Do you have an emergency plan and, where necessary, have you recorded the details? ☐

- Does your plan take account of other emergency plans applicable to the same buiiding? ☐

- Is the plan readily available for staff to read? ☐

- Is the emergency plan available to the enforcing authority? ☐

4.3   Inform, instruct, co-operate and co-ordinate

You must give clear and relevant information and appropriate instructions to your staff and the employers of other people working in your premises, such as contractors, about how to prevent fires and what they should do if there is a fire.

If you intend to employ a child, you must inform the parents of the significant risks you have identified and the precautions you have taken. You must also co-operate and co-ordinate with other responsible people who use any part of the premises. It is unlikely that your emergency plan will work without this.

Information and instruction

All staff should be given information and instruction as soon as possible after they are appointed and regularly after that. Make sure you include staff who work outside normal working hours, such as contract cleaners or maintenance staff.

All other relevant persons should be given information about the fire safety arrangements as soon as possible. Information should be available for the public, e.g. fire action notices.

The information and instructions you give must be in a form that can be used and understood. They should take account of those with disabilities such as hearing or sight impairment, those with learning difficulties and those who do not use English as their first language.

The information and instruction you give should be based on your emergency plan and must include:

- the significant findings from your fire risk assessment;

- the measures that you have put in place to reduce the risk;

- what staff should do if there is a fire;

- the identity of people you have nominated with responsibilities for fire safety; and

- any special arrangements for serious and imminent danger to persons from fire.

Information and instruction may simply involve an explanation of the fire procedures and how they are to be applied. This should include showing staff the fire-protection arrangements, including the designated escape routes, the location and operation of the fire-warning system and any other fire-safety equipment provided, such as fire extinguishers. Fire action notices can complement this information and, where used, should be posted in prominent locations.

Further guidance on information and instruction to staff, and on working with dangerous substances, is given in Part 2, Section 7.3.

Co-operation and co-ordination

In premises that are not multi-occupied you are likely to be solely responsible. However, in buildings owned by someone else, or where there is more than one occupier, and others are responsible for different parts of the building, it is important that you liaise with them and inform them of any significant risks that you have identified. By liaising you can co-ordinate your resources to ensure that your actions and working practices do not place others at risk if there is a fire, and a co-ordinated emergency plan operates effectively.

Where two or more responsible persons share premises in which an explosive atmosphere may occur, the responsible person with overall responsibility for the premises must co-ordinate any measures necessary to protect everyone from any risk that may arise. Employees also have a responsibility to co-operate with their employer so far as it is necessary to help the employer comply with any legal duty.

Further guidance on co-operation and co-ordination is given in Part 2, Section 7.3.

Checklist

- Have you told your staff about the emergency plan? ☐

- Do you know the location and identity of people you have nominated to do a particular task? ☐

- Have you given staff information about any dangerous substances? ☐

- Do you have arrangements for informing temporary or agency staff? ☐

- Do you have arrangements for informing other employers whose staff are guest workers in your premises, such as maintenance contractors and cleaners? ☐

- Have you informed guests and visitors about what to do in an emergency? ☐

- Have you co-ordinated your fire safety arrangements with other responsible people and with any contractors in the building? ☐

- Have you recorded details of any information or instructions you have given and the details of any arrangements for co-operation and co-ordination with others? ☐

4.4 Fire safety training

You must provide adequate fire safety training for your staff. The type of training should be based on the particular features of your premises and should:

- take account of the findings of the fire risk assessment;

- explain your emergency procedures;

- take account of the work activity and explain the duties and responsibilities of staff;

- take place during normal working hours and be repeated periodically where appropriate;

- be easily understandable by your staff and other people who may be present; and

- be tested by fire drills.

In simple premises this may be no more than showing new staff the fire exits and giving basic training on what to do if there is a fire. In most premises, in particular where there is a high staff turnover and many shift patterns, the organisation of fire safety training will need to be more formal, e.g. by an induction course. Some of your staff may have received some fire safety training as part of a national accreditation scheme (e.g. door staff and stewards).

Your training should include the following:

- what to do on discovering a fire;

- how to raise the alarm and what happens then;

- what to do upon hearing the fire alarm;

- the procedures for alerting members of the public and visitors including, where appropriate, directing them to exits;

- the arrangements for calling the fire and rescue service;

- the evacuation procedures for everyone (including the public and contractors) in your premises to reach an assembly point at a place of total safety;

- the location and, when appropriate, the use of firefighting equipment;

- the location of escape routes, especially those not in regular use;

- how to open all emergency exit doors;

- the importance of keeping fire doors closed to prevent the spread of fire, heat and smoke;

- where appropriate, how to stop machines and processes and isolate power supplies in the event of a fire;

- the reason for not using lifts (except those specifically installed or nominated, following a suitable fire risk assessment;

- the safe use of and risks from storing or working with highly flammable and explosive substances; and

- the importance of general fire safety, which includes good housekeeping.

All the staff identified in your emergency plan that have a supervisory role if there is a fire (e.g. fire marshals or wardens), should be given details of your fire risk assessment and receive additional training.

Further guidance on training and how to carry out a fire drill is given in Part 2, Section 7.4.

Checklist

- Have your staff received any fire safety training? ☐
- Have you carried out a fire drill recently? ☐
- Are employees aware of specific tasks if there is a fire? ☐
- Are you maintaining a record of training sessions? ☐
- Do you carry out joint training and fire drills in multi-occupied buildings? ☐
- If you use or store hazardous or explosive substances have your staff received appropriate training? ☐

STEP 5   REVIEW

You should constantly monitor what you are doing to implement the fire risk assessment, to assess how effectively the risk is being controlled.

If you have any reason to suspect that your fire risk assessment is no longer valid or there has been a significant change in your premises that has affected your fire precautions, you will need to review your assessment and if necessary revise it. Reasons for review could include:

- changes to work activities or the way that you organise them, including the introduction of new equipment;

- alterations to the building, including the internal layout;

- substantial changes to furniture and fixings;

- the introduction, change of use or increase in the storage of hazardous substances;

- the failure of fire precautions, e.g. fire-detection systems and alarm systems, life safety sprinklers or ventilation systems;

- significant changes to displays or quantities of stock;

- a significant increase in the number of people present; and

- the presence of people with some form of disability.

You should consider the potential risk of any significant change before it is introduced. It is usually more effective to minimise a risk by, for example, ensuring adequate, appropriate storage space for an item before introducing it to your premises.

Do not amend your assessment for every trivial change, but if a change introduces new hazards you should consider them and, if significant, do whatever you need to do to keep the risks under control. In any case you should keep your assessment under review to make sure that the precautions are still working effectively. You may want to re-examine the fire prevention and protection measures at the same time as your health and safety assessment.

If a fire or 'near miss' occurs, this could indicate that your existing assessment may be inadequate and you should carry out a re-assessment. It is good practice to identify the cause of any incident and then review and, if necessary, revise your fire risk assessment in the light of this.

Records of testing, maintenance and training, etc. are useful aids in a review process. See Appendix A1 for an example.

Alterations notices

If you have been served with an 'alterations notice' check it to see whether you need to notify the enforcing authority about any changes you propose to make as a result of your review. If these changes include building work, you should also consult a building control body.

END OF PART 1

You should now have completed the five-step fire risk assessment process, using the additional information in Part 2 where necessary. In any review you may need to revisit Steps 1 to 4.

Part 2 Further guidance on fire risk assessment and fire precautions

Managing fire safety

Good management of fire safety in your premises is essential to ensure that any fire safety matters that arise are always effectively addressed. In simple premises this can be achieved by the manager or owner maintaining and planning fire safety in conjunction with general health and safety.

In larger premises it is good practice for a senior manager to have overall responsibility for fire safety. It may be appropriate for this responsibility to be placed with the manager designated with overall responsibility for health and safety.

If your premises are owned or managed by a corporate organisation, e.g. a national company with multiple outlets, the organisation's fire safety policy should be flexible enough to allow modification. This is particularly important when local managers have to function daily with other businesses in the same building. It should be recognised that fire safety operates at all levels within an organisation and therefore local managers should be able to develop, where necessary, a local action plan for their premises.

The company policy should be set out in writing and may cover such things as:

- who will hold the responsibility for fire safety at board level;

- who will be the responsible person for each of their premises (this will be the person who has overall control, usually the manager, but may be part-time or shift managers);

- the arrangement whereby those responsible for fire safety will, where necessary, nominate in writing specific people to carry out particular tasks if there is a fire; and

- the arrangement whereby regional or area managers should monitor and check that individual managers are meeting the requirements of the fire safety law.

You should have a plan of action to bring together all the features you have evaluated and noted from your fire risk assessment so that you can logically plan what needs to be done. It should not be confused with the emergency plan, which is a statement of what you will do if there is a fire.

The plan of action should include what you intend to do to reduce the hazards and risks you have identified and to implement the necessary protection measures.

You will need to prioritise these actions to ensure that any findings which identify people in immediate danger are dealt with straight away, e.g. unlocking fire exits. In other cases where people are not in immediate danger but action is still necessary, it may be acceptable to plan this over a period of time.

You need to be aware that certain functions e.g. discos, can present additional dangers for the audience, largely from the effects of over excitement and irrational behaviour; together with the higher noise level and flashing lights. In such circumstances, and particularly where there is a mainly younger audience, you should ensure that there are a sufficient number of competent and adequately trained attendants to cover an emergency situation; and a public address system which can over-ride the performance and be heard clearly in all parts of the premises will be required.

Before admitting the public to your premises you need to ensure that all of your fire safety provisions are in place and in working order, or, if not, that alternative arrangements are in place. Constant checks are needed while the public are present, and again after they have left. Detailed recommendations are given in BS5588-12.[51]

It is important that the manager ensures that where, for operational reasons, staff are transferred to another part of the premises, they are made aware of the means of escape and fire procedures of the new work area (including any additional responsibilities) if this differs from their permanent duty station.

Where your premises or parts of your premises are either hired or leased then the management responsibilities of the hirer should be defined.

For a place of assembly, the fire and rescue service should be called to every outbreak of fire.

The guidance in Part 2 provides additional information to:

- ensure good fire safety management by helping you establish your fire prevention measures, fire precautions and fire safety procedures (systems, equipment and plans); and

- assist you to carry out your fire safety risk assessment and identify any issues that need attention.

Section 1 Further guidance on fire risks and preventative measures

This section provides further information on evaluating the risk of a fire and its prevention in your premises. You should spend time developing long-term workable and effective strategies to reduce hazards and the risk of a fire starting. At its simplest this means separating flammable materials from ignition sources.

You should consider:

- housekeeping;
- storage;
- dangerous substances: storage and use;
- equipment and machinery;
- electrical safety;
- smoking;
- managing building work and alterations;
- existing layout and construction;
- particular hazards in corridors and stairways used as escape routes;
- insulated core panels;
- restricting the spread of fire and smoke;
- arson; and
- help for people with special needs.

1.1 Housekeeping

Good housekeeping will lower the chances of a fire starting, so the accumulation of combustible materials in premises should be monitored carefully. Good housekeeping is essential to reduce the chances of escape routes and fire doors being blocked or obstructed.

Figure 13: Bins under a stairway (courtesy of Cheshire fire and rescue service)

Refuse

The accumulation of waste and litter (such as programmes and food and drink packaging) should be avoided. All parts of the premises should therefore be inspected before, during (as far as is practicable) and after each event or period open to the public. Any accumulation of waste or litter should be removed without delay or kept in a fire-resisting container or room, pending removal.

Sufficient waste and litter bins should be provided and arrangements made for their frequent emptying, especially during an event or function of extended duration.

Waste material should be kept in suitable containers prior to removal from the premises. If bins, particularly wheeled bins, are used outside, they should be secured in a compound to prevent them being moved to a position next to the building and set on fire. Skips should never be placed against a building and should normally be a minimum of 6m away from any part of the premises (see Figure 13).

If you generate a considerable quantity of combustible waste material then you may need to develop a formal plan to manage this effectively.

In high risk areas you need to make sure arrangements are in place for safe close down, e.g. checking all appliances are turned off and combustible waste has been removed.

1.2 Storage

Many of the materials found in your premises will be combustible. If your premises have inadequate or poorly managed storage areas then the risk of fire is likely to be increased (see Figure 14). The more combustible materials you store the greater the source of fuel for a fire. Poorly aranged storage could prevent equipment such as sprinklers working effectively.

Combustible materials are not just those generally regarded as highly combustible, such as polystyrene, but all materials that will readily catch fire. However, by carefully considering the type of material, the quantities kept and the storage arrangements, the risks can be significantly reduced.

Figure 14: An example of poor storage

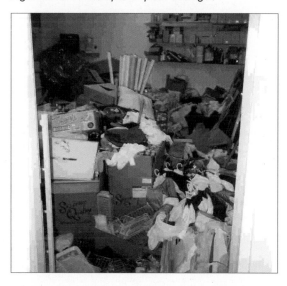

Stacks of combustible materials (e.g. seating or, in offices, the retention of large quantities of paper records) can increase the fire hazard. Such readily available combustible material makes the potential effect of arson more serious.

Case study

A church owned a petrol-driven lawnmower. When not in use, the lawnmower was stored under the stairs inside the church, and exposed to the public. The lawnmower presented a significant risk, and could be a target for arson.

The lawnmower should have been stored in a properly ventilated fire-resisting storeroom away from the church.

Many premises will take great care to present an efficient and attractive image in the public areas, while storage areas are neglected and allowed to become dumping areas for unused material.

Do not pile combustible material against electrical equipment or heaters, even if turned off for the summer, and do not allow smoking in areas where combustible materials are stored.

To reduce the risk, store excess materials in a dedicated storage area, storeroom or cupboard. Do not store excess materials in areas where the public would normally have access.

Consider how materials are stored and evaluate any additional risk of fire that it generates. For example, to reduce the potential rate of fire growth, when not in use foam gym mats should be laid flat and stacked on top of one another in a dedicated storeroom. Stacked side by side or vertically against a wall, the potential speed with which the fire may grow is significantly increased.

Your fire risk assessment should also consider any additional risk generated by seasonal products such as fireworks and Christmas decorations.

Consider the following to reduce these risks:

- ensure storage and display areas are adequately controlled and monitored;

- use fire-retardant display materials wherever possible (suppliers should be able to provide evidence of this); and

- ensure electrical lighting used as part of the display does not become a potential source of ignition.

Furnishings, upholstered seating and cushions
When stored in bulk, certain types of cushioning (e.g. which is distributed or hired for event use), foam mats and ball pit play areas pose a risk of a rapid fire growth and should therefore be stored in a fire-resisting container or room. (See also Section 1.11.)

Cellars and basements
Cellars and basements are often used for storage, of items such as stock and promotional materials, fuel used for outdoor barbeques and patio heaters. Care may be needed if the cellar connects directly to the rest of the building, especially if the cellar is accessible to the public. Cellar areas may also offer a target for arsonists and should be secured against unauthorised access.

Scenery store, storage enclosures, open stage storage

Because scenery often comprises combustible materials, you need to take particular care with its storage and, in particular, storage on an open stage.

Only materials which you know are not combustible should be stored on an open stage; otherwise such materials should be stored in a 30-minute fire-resisting storage enclosure when they are not in current use. If in doubt you should seek specialist advice.

1.3 Dangerous substances: storage, display and use

Specific precautions are required when handling and storing dangerous substances to minimise the possibility of an incident. Your supplier should be able to provide detailed advice on safe storage and handling; however, the following principles will help you reduce the risk from fire:

- substitute highly flammable substances and materials with less flammable ones;

- reduce the quantity of dangerous substances to the smallest reasonable amount necessary for running the business or organisation;

- correctly store dangerous substances, e.g. in a fire-resisting enclosure. All flammable liquids and gases should ideally be locked away, especially when the premises are unoccupied, to reduce the chance of them being used in an arson attack; and

- ensure that you and your employees are aware of the fire risk the dangerous substances present and the precautions necessary to avoid danger.

Additional general fire precautions may be needed to take account of the additional risks that may be posed by the storage and use of these substances.

Certain substances and materials are by their nature, highly flammable, oxidising or potentially explosive. These substances are controlled by other legislation in addition to fire safety law, in particular the Dangerous Substances and Explosive Atmospheres Regulations 2002[7] (also see the HSE's *Approved code of practice and guidance*[8]).

Flammable liquids

Highly flammable liquids present a particularly high fire risk. For example, a leak from a container of flammable solvents, such as acetone, may produce large quantities of heavier-than-air flammable vapours. These can travel large distances, increasing the likelihood of their reaching a source of ignition well away from the original leak, such as a basement containing heating plant and/or electrical equipment on automatic timers.

Flammable liquids stored in plastic containers can be a particular problem if involved in fire because they readily melt, spilling their contents and fuelling rapid fire growth.

The risk is reduced by ensuring the storage and use of highly flammable liquids is carefully managed, that materials contaminated with solvent are properly disposed of (see Figure 15) and when not in use, they are safely stored. Up to 50 litres may be stored in a fire-resisting cabinet or bin that will contain any leaks (see Figure 16).

Figure 15: A fire-resisting pedal bin for rags

Figure 16: A 50-litre flammable storage bin

There should be no potential ignition sources in areas where flammable liquids are used or stored and flammable concentrations of vapours may be present. Any electrical equipment used in these areas, including fire alarm and emergency lighting systems, needs to be suitable for use in flammable atmospheres. In such situations, it is recommended that you seek advice from a competent person.

LPG storage and use

Where LPG in cylinders or cartridges is present, you need to take particular care to minimise the possibility of its involvement in a fire. The total amount of LPG should be kept to the minimum necessary to meet your needs.

Where it is necessary to keep spare LPG cylinders in the place of assembly, this should be limited to 15kg in not more than two cylinders. These should be kept in a safe, secure and well ventilated place where:

- they cannot be interfered with;

- they can be kept upright (with valve protection fitted);

- they are away from sources of ignition and/or ignitable materials; and

- are away from any corrosive toxic or oxidant materials.

Where there are retail premises within the place of assembly, holding LPG cylinders and cartridges, the maximum stock should not exceed 70kg, which should be kept in a safe place, where they cannot be interfered with, and away from stairways, exit doors and places where they might obstruct means of escape.

Stock in excess of 70kg should be kept in a properly designed and located storage area away from the premises or in a building that is dedicated for LPG. Advice on the use of LPG for heating is given in Section 1.4.

Bulk storage tanks for LPG should be designed, installed and located in accordance with industry guidance.

Further guidance on the safe storage of LPG is available from your supplier or the Liquefied Petroleum Gas Association's Code of Practice.[9]

Piping

Piping conveying gas or flammable liquid should be, as far as practicable, of rigid metal. Any necessary flexible piping should consist of material suitable for the gas or liquid being conveyed; it should be adequately reinforced to resist crushing and withstand the maximum internal pressure to which it may be subjected. Any connections to the flexible piping should be of an approved pattern (i.e. screwed or otherwise secured with a patent design device to prevent accidental disconnection). If in doubt you should seek advice from a competent person.

Storage and use of cylinders

Only those compressed gas cylinders or gas vessels containing flammable liquids or gas under pressure required for immediate use should be kept within the premises. Any spare cylinders should be stored in a safe and secure location outside the building.

Hazardous materials

Where it is necessary to utilise hazardous materials, such as fuels (whether in containers or within fuel tanks and machinery), fertilisers, weed killers, paints, they should be stored in a secure and safe location, for example a properly ventilated fire-resistant storeroom. Care should also be taken to ensure that incompatible materials are not stored together. If in any doubt you should seek the advice of your supplier.

Explosives and flammable substances

You should adopt and enforce a clear policy prohibiting people from bringing flares or fireworks into the premises.

Explosive or highly flammable substances should only be brought onto the premises under specified and monitored conditions.

All pyrotechnics and similar substances must be operated safely and in accordance with manufacturer's instructions. 'Homemade' pyrotechnics or those with a doubtful origin should not be used.

Pyrotechnics intended for use in the open air should not be used inside your premises.

Pyrotechnics should only be used by a person(s) who is trained and competent to handle them.

Issues that need to be considered include:

- siting of the pyrotechnic devices;

- type of firing device to be employed;

- use of bomb tanks;

- storage, storage receptacles and enclosures; and

- general safety and security.

Detailed recommendations on the use and storage of explosives and flammable substances are given by the Association of British Theatre Technicians[52] and the Home Office/Scottish Office.[49]

Real flame

Real flame (including candles) should not be used in an entertainment unless your risk assessment shows that it can be done safely. Where real flame is used, a trained member of your staff should monitor the activity and have an appropriate fire extinguisher readily available.

Smoke and vapour effects, lasers and strobes and firearms

Such equipment should be operated safely and in accordance with manufacturer's instructions. The recommendations regarding the use of pyrotechnics (see above), equipment and machinery and electrical safety should be applied as appropriate. However, such devices can create an illusion of the presence of fire (and/or smoke) and consequently some additional measures may need to be taken to avoid unnecessary alarm and also to ensure that the use of such devices does not disguise the presence of a real fire.

Detailed recommendations on the use of smoke and vapour effects, lasers and strobes and firearms are given by the HSE.[51]

Premises licensed under the Licensing Act 2003[47] may have conditions prohibiting or restricting the use of explosives, flammable substances, real flame, smoke and vapour effects, lasers and strobes or firearms. You may need to consult the local licensing authority on this matter.

1.4 Equipment and machinery

Common causes of fire in equipment and machinery are:

- allowing ventilation points to become clogged or blocked, causing overheating;

- allowing extraction equipment in catering environments to build up excessive grease deposits;

- loose drive belts or lack of lubrication leading to increased friction;

- disabling or interfering with automatic or manual safety features and cut-outs; and

- leaking valves, glands or joints allowing oils and other flammable liquids to contaminate adjacent floors or goods.

All machinery, equipment and plant, including cooking and heating equipment, and office equipment such as photocopiers, should be properly maintained by a competent person. Appropriate signs and instructions on safe use may be necessary.

Heating

Individual heating appliances require particular care if they are to be used safely, particularly those which are kept for emergency use during a power cut or as supplementary heating during severe weather. The greatest risks arise from lack of maintenance and staff unfamiliarity with them. Heaters should be secured in position when in use and fitted with a fire guard if appropriate.

As a general rule, convector or fan heaters should be preferred to radiant heaters because they present a lower risk of fire and injury. The following rules should be observed:

- All heaters should be kept well clear of combustible materials and where they do not cause an obstruction.

- Heaters which burn a fuel should be sited away from draughts.

- Portable fuel burning heaters (including bottled gas (LPG)) should only be used in public areas in exceptional circumstances and if shown to be acceptable in your risk assessment.

All gas heating appliances should be used only in accordance with manufacturer's instructions and should be serviced annually by a competent person.

1.5 Electrical safety

Electrical equipment is a significant cause of accidental fires in premises. The main causes are:

- overheating cables and equipment, e.g. due to overloading circuits, bunched or coiled cables or impaired cooling fans;

- incorrect installation or use of equipment;

- damaged or inadequate insulation on cables or wiring;

- combustible materials being placed too close to electrical equipment which may give off heat even when operating normally or may become hot due to a fault;

- arcing or sparking by electrical equipment; and

- lack of maintenance or testing.

All electrical equipment should be installed and maintained in a safe manner by a competent person. If portable electrical equipment is used, including items brought into a workplace by staff, then your fire risk assessment should ensure that it is visually inspected and undergoes portable appliance testing ('PAT') at intervals suitable for the type of equipment and its frequency of use (refer to HSE guidance[10]). If you have any doubt about the safety of your electrical installation then you should consult a competent electrician.

Issues to consider include:

- overloading of equipment;

- correct fuse ratings;

- PAT testing and testing of the fixed installation;

- protection against overloading of installation;

- protection against short circuit;

- insulation, earthing and electrical isolation requirements;

- frequency of electrical inspection and test;

- annual electrical inspection and test;

- temperature rating and mechanical strength of flexible cables;

- portable electrical equipment, including lanterns;

- physical environment in which the equipment is used (e.g. wet or dusty atmospheres); and

- suitable use and maintenance of personal protective equipment.

All electrical installations should be regularly maintained by a competent electrical engineer, appointed by you, or on your behalf, in accordance with the Electricity at Work Regulations 1989 (EAW Regulations).[55] The use of low voltage equipment should conform to the requirements of the Electrical Equipment (Safety) Regulations 1994,[56] including the requirement to be CE marked.

1.6 Smoking

Carelessly discarded cigarettes and other smoking materials are a major cause of fire. A cigarette can smoulder for several hours, especially when surrounded by combustible material. Many fires are started several hours after the smoking materials have been emptied into waste bags and left for future disposal.

If it is currently allowed in your premises, consider prohibiting smoking other than in the designated smoking areas. Display suitable signs throughout the premises informing people of the smoking policy and the locations where smoking is permitted. Smoking should be prohibited in those parts of the premises which are constructed of, or contain, combustible or flammable items or materials.

In those areas where smoking is permitted, provide deep and substantial metal ashtrays to help prevent unsuitable containers being used. Empty all ashtrays daily into a metal waste bin and take it outside. It is dangerous to empty ashtrays into plastic waste sacks which are then left inside for disposal later.

1.7 Managing building work and alterations

Fires are more frequent when buildings are undergoing refurbishment or alteration.

You should ensure that, before any building work starts, you have reviewed the fire risk assessment and considered what additional dangers are likely to be introduced. You will need to evaluate the additional risks to people, particularly in those buildings that continue to be occupied. Lack of pre-planning can lead to haphazard co-ordination of fire safety measures.

You should liaise and exchange information with contractors who will also have a duty under the Construction (Health, Safety and Welfare) Regulations 1996[11, 12] to carry out a risk assessment and inform you of their significant findings and the preventive measures they may employ. This may be supported by the contractors' agreed work method statement. The designer should also have considered fire safety as part of the Construction (Design and Management) Regulations 1994 (the CDM Regulations).[57]

You should continuously monitor the impact of the building work on the general fire safety precautions, such as the increased risk from quantities of combustible materials and accumulated waste and maintaining adequate means of escape. You should only allow the minimum materials necessary for the work in hand within or adjacent to your building.

Activities involving hot work such as welding, flame cutting, use of blow lamps or portable grinding equipment can pose a serious fire hazard and need to be strictly controlled when carried out in areas near flammable materials. This can be done by having a written permit to work for the people involved (whether they are your employees or those of contractor).

A permit to work is appropriate in situations of high hazard/risk and, for example, where there is a need to:

- ensure that there is a formal check confirming that a safe system of work is being followed;

- co-ordinate with other people or activities;

- provide time limits when it is safe to carry out the work; and

- provide specialised personal protective equipment (such as breathing appratus) or methods of communication.

Additional risks that can occur during building work include:

- hot work such as flame cutting, welding or, soldering, paint stripping;

- temporary electrical equipment;

- blocking of escape routes, including external escape routes;

- introduction of combustibles into an escape route;

- loss of normal storage facilities;

- fire safety equipment, such as automatic fire-detection systems becoming affected;

- fire-resisting partitions being breached or fire doors being wedged open (see Appendix B1 for information on fire-resisting separation); and

- additional personnel who may be unfamiliar with the premises.

You must notify the fire and rescue service about alterations in your premises if an alterations notice is in force.

Further guidance on fire safety during construction work is available from the HSE[58, 59] and the Fire Protection Association.[60]

1.8 Existing layout and construction

In many premises, the design is for open-plan areas allowing members of the public and staff to move freely throughout the floor.

Traditionally, occupants are advised to shut doors when escaping from a fire but in open-plan areas this may not be possible. In these areas the fire, and especially the smoke, may spread faster than expected.

To assess the risk in your premises you need to evaluate the construction and layout of your premises. This does not mean a structural survey, unless you suspect that the structure is damaged or any structural fire protection is missing or damaged, but rather an informed look around to see if there are any easy paths through which smoke and fire may spread and what you can do to stop that. In general, older buildings will have more void areas, possibly hidden from view, which will allow smoke and fire to spread away from its source. Whatever your type of building, you may need to consider typical situations that may assist the spread of fire and smoke such as:

- vertical shafts, e.g. lifts, open stairways, dumb waiters or holes for moving stock around;

- false ceilings, especially if they are not fire-stopped above walls;

- voids behind wall panelling;

- unsealed holes in walls and ceilings where pipe work, cables or other services have been installed; and

- doors, particularly to stairways, which are ill-fitting or routinely left open.

Marquees, tents and temporary structures
Marquees, tents and temporary structures should be of proven fire performance. Any flexible membrane covering a structure (other than an air supported structure) should comply with the recommendations given in Appendix A of BS 7157.[61] Air supported structures should comply with the recommendations given in BS 6661.[62]

Emergency lighting should be provided on escape routes.

Ensure that all long grass around a marquee is cut before the marquee is erected and remove the cuttings to prevent the risk of fire.

1.9 Particular hazards in escape routes

Items that are a source of fuel, pose an ignition risk, or are combustible and likely to increase the fire loading or spread of fire, should not be located on any corridor or stairway or circulation space that will be used as an escape route. Such items include:

- portable heaters, e.g. bottled gas (LPG) or electric radiant heaters and electric convectors or boilers;

- gas cylinders for supplying heaters;

- cooking appliances; and

- unenclosed gas pipes, meters, and other fittings.

However, depending on the findings of your risk assessment and where more than one escape route is available, the items below may be acceptable if the minimum exit widths are maintained and the item presents a relatively low fire risk:

- coat racks;

- non-combustible lockers;

- vending machines;

- small items of electrical equipment (e.g. photocopiers); and

- small quantities of upholstered furniture which meets BS 7176[83] or the furniture and furnishing (Fire) (Safety) Regulations 1988.[64]

1.10 Insulated core panels

Many buildings have insulated core panels as exterior cladding or for internal structures and partitions. They are often used because they are easily constructed, which enables alterations and additional internal partitions to be erected with minimum disruption to business.

They normally consist of a central insulated core, sandwiched between an inner and outer metal skin (see Figure 17). The central core can be made of various insulating materials, ranging from virtually non-combustible through to highly combustible. Fire can grow unnoticed in the core and only become apparent when it is well developed.

Figure 17: Insulated core panels – internal panel

It is difficult to identifty the type of core the panels have, therefore best practice can help you reduce any additional risk.

- Do not store highly combustible materials, or install heating appliances, such as baking ovens, against the panels.

- Control ignition sources that are adjacent to, or penetrating the panels.

- Have damaged panels or sealed joints repaired immediately and make sure that jointing compounds or gaskets used around the edges of the panels are in good order.

- Check where openings have been made for doors, windows, cables and ducts to ensure that these have been effectively sealed and the inner core has not been exposed.

- Check that there has been no mechanical damage, e.g. caused by mobile equipment.

- Ensure that any loads, such as storage and equipment, are only supported by panels that have been designed and installed to perform this function.

- Check that inner and outer skins are adhering tightly to the core.

The use of combustible panels in areas of buildings with a high life risk, e.g. where large numbers of people are present, should be carefully considered. Your fire risk assessment may need to be revised to ensure that any increased risk resulting from this type of construction is considered.

The panels should be installed by a competent person in accordance with industry guidance.

Guidance on the design, construction, specification and fire management of insulated core panels has been published by the International Association for Cold Storage Construction.[63]

1.11 Restricting the spread of fire and smoke

To reduce the risk to people if there is a fire, you need to consider how to control or restrict the spread of fire and smoke. The majority of people who die in fires are overcome by the smoke and gases.

It is important ensure that, in the event of fire, the rate of fire growth is restricted in its early stages. It should also be noted that most measures which restrict the rate of fire growth in its early stages will also serve to restrict the fire spread in its later stages.

Catering facilities

Wherever possible, any extensive catering facility, particularly those with deep fat fryers, should be located in separate buildings. If located within other buildings they should be separated from the rest of the building by fire-resisting construction and provided with adequate ventilation. Where flues pass through any part of the structure, the structure should be protected by fire-resisting construction and the flue should terminate at a point where emissions can disperse in the open air. Where fire shutters are used these should be capable of operating both manually and by fusible link. Where a fire detection and warning system is installed the fire shutter should also be designed to close on activation of the system via a controlled geared mechanism.

Combustible contents

Your premises will contain a range of combustible contents. These are likely to include:

- furnishings, upholstered seating, furniture and cushions;

- curtains, drapes and other textile hangings;

- artificial and dried foliage;

- decorative textile fabrics (e.g. within stands);

- sports and play area furnishings, such as cellular foam gym mats; and

- scenery or properties used for stage presentations.

The use of furnishings and other materials which are easily ignited or have rapid spread of flame characteristics should be avoided you should seek to use materials that are either non combustible of durably or inherently flame-retardant. Any fabrics used in escape routes, other than foyers, entertainment areas or function rooms should be non combustible. Drapes and curtains should not be provided across escape routes or exits.

You should note that materials treated with flame-retardant treatments may have a limited 'wash life' before the effectiveness of the flame-retardant is diminshed. To maintain the protection you should follow the manufacturers/suppliers instructions.

Similarly, scenery and properties used for stage presentations often comprise combustible materials, so you need to take particular care with their use and, in particular, when they are on an open stage you should seek to only use materials which you know are not combustible.

Upholstered seating, carpets and other textile floor coverings and underlays should be resistant to ignition.

Gymnastic mats and similar equipment should ideally contain only combustion modified high resilience foam.

Specifications for the fire performance and testing of furniture and furnishings can be found in the following standards:

- ignitability of upholstered furniture (including composites of cover material and infill) and loose covers: BS 5852;[82]

- resistance to ignition of upholstered furniture for non-domestic use: BS 7176;[83]

- flammability of fabrics for curtains and drapes (including nets and linings): BS 5867-2;[66]

- burning behaviour (ignitability and flame spread) of curtains and drapes: BS EN 1101[84] (ignitability and BS EN 1102[85] (flame spread);

- assessment and labelling of textile floor coverings: BS 5287;[90] and

- gymnasium eqiupment: BS 1892-2.[67]

Equivalent European classification standards may also be used.

If in doubt you should seek specialist advice with regards to the treatments and tests for these materials which can reduce their flammability and/or combustibility.

Display materials and decorations

You should evaluate what material could ignite first and what would cause the fire to develop and spread, and assess how materials used in temporary or permanent displays would interact with surface linings and position them accordingly. In particular, displays such as paper, textiles, Christmas decorations and trees or other flimsy materials should not be located in stairways or corridors. However, such materials may be acceptable in other locations if treated with an appropriate fire-retardant product.

Staff information should be confined to appropriately located display boards in areas away from escape routes. Display boards may be used on escape routes as long as they are no bigger than 1m² or have been enclosed in a sealed display case.

Ad-hoc flame tests
Ad-hoc flame tests can be used to screen out some materials. Simple match tests, or using a cigarette lighter, can be used to assess ease of ignition, but any such tests must be carried out in the open air, away from combustible materials, and with a fire extinguisher to hand.

Fire-resisting structures
Many buildings are divided into different areas by fire doors and fire-resisting walls and floors. These are partly designed to keep a fire within one area, giving people more time to escape. You will need to identify which doors, walls and floors in your building are fire-resisting. There may be information available from when the building was built, if alterations have been made, or from a previously held fire certificate.

High-risk areas (e.g. extensive catering facilities) should be seperated from the rest of the premises by 30-minute fire-resisting construction.

Normally if there are fire doors in a wall, then the wall itself will also need to be fire-resisting (see Appendix B for technical information about fire-resisting walls and doors). If a wall or floor is required to be fire-resisting then you should not make any holes in it, e.g. for extra doors or pipe ducts, without consulting a competent person.

Ventilation systems
Where ventilation systems might assist the spread of flames, smoke and hot gases from a fire it will be necessary to take steps to safeguard the means of escape against this hazard.

1.12 Arson

Recent studies indicate that, across all premises types, over 2,100 deliberately set fires, resulting in two deaths and 55 injuries, occur every week. All premises can be targeted either deliberately or just because they offer easy access.

Be aware of other small, deliberately set fires in the locality, which can indicate an increased risk to your premises. Be suspicious of any small 'accidental' fires on the premises and investigate them fully and record your findings.

Fires started deliberately can be particularly dangerous because they generally develop much faster and may be intentionally started in escape routes. Of all the risk-reduction measures, the most benefit may come from efforts to reduce the threat from arson.

Measures to reduce arson may include the following:

- ensure the outside of the premises is well lit and, if practical, secure the perimeter of the premises;

- thoroughly secure all entry points to the premises, including windows and the roof, but make sure that this does not compromise people's ability to use the escape routes;

- make sure you regularly remove all combustible rubbish;

- do not place rubbish skips adjacent to the building and secure waste bins in a compound separated from the building;

- do not place vehicles, caravans or other portable structures adjacent to the building;

- encourage staff to report people acting suspiciously;

- remove automatic entry rights from staff who have been dismissed;

- ensure that your security alarm/fire-detection system is monitored and acted on;

- secure flammable liquids so that intruders cannot use them;

- secure all storage areas and unused areas of the building that do not form part of an escape route against unauthorised access, ensure access to keys to those areas is restricted; and

- fit secure metal letterboxes on the inside of letter flaps to contain any burning materials that may be pushed through and ensure regular inspection for combustible material or sources of ignition.

Further guidance on reducing the risk of arson has been published by the Arson Prevention Bureau.*

1.13 Help for people with special needs

Of all the people who may be especially at risk you will need to pay particular attention to people who have special needs including those with a disability. The Disability Rights Commission estimates that 11 million people in this country have some form of disability, which may mean that they find it more difficult to leave a building if there is a fire. Under the Disability Discrimination Act,[13] if disabled people could realistically expect to use premises, then you must anticipate any reasonable adjustments that would make it easier for that right to be exercised.

The Disability Discrimination Act[13] includes the concept of 'reasonable adjustments' and this can be carried over into fire safety law. It can mean different things in different circumstances. For a small business it may be considered reasonable to provide contrasting colours on a handrail to help people with vision impairment to follow an escape route more easily. However, it might be unreasonable to expect that same business to install an expensive voice-alarm system.

If disabled people are going to be in your premises then you must also provide a safe means for them to leave if there is a fire. You and your staff should be aware that disabled people may not react, or can react differently, to a fire warning or fire. You should give similar consideration to others with special needs such as parents with young children or the elderly.

In premises with a simple layout, a common-sense approach, such as offering to help lead a blind person or helping an elderly person down steps may be enough. In more complex premises, more elaborate plans and procedures will be needed, with trained staff assigned to specified duties. In complex premises, you may also wish to contact a professional consultant or take advice from disability organisations.

Whilst the majority of people with special needs wish to and are able to make their own escape, there may be a number who are only able to move or react adequately with assistance from staff.

Consider the needs of those with mental disabilities or spatial recognition problems. The range of disabilities encountered can be considerable, extending from mild epilepsy to complete disorientation in an emergency situation. Many of these can be addressed by properly trained staff, discreet and empathetic use of the 'buddy system' or by careful planning of colour and texture to identify escape routes.

Where people with special needs use or work in the premises, their needs should, so far as is practicable, be discussed with them. These will often be modest and may require only changes or modifications to existing procedures. You may need to develop individual 'personal emergency evacuation plans' (PEEPs) for disabled persons who frequently use a building. They will need to be confident of any plan/PEEP that is put in place after consultation with them. As part of your consultation exercise you will need to consider the matter of personal dignity.

If members of the public use your building then you may need to develop a range of standard PEEPs which can be provided on request to a disabled person or others with special needs.

Guidance on removing barriers to the everyday needs of disabled people is in BS 8300.[14] Much of this advice will also help disabled people during an evacuation.

You should also consider the particular needs of children and the elderly who may be using your premises. Similarly, special consideration may be needed if members of the public are likely to be intoxicated, or where the activity undertaken may limit the ability, or readiness, to escape (e.g. swimming pools).

Further advice can be obtained from the Disability Rights Commission at www.drc-gb.org

---

* Visit www.arsonpreventionbureau.org.uk for more information.

Section 2 Further guidance on fire detection and warning systems

Where a simple shout of 'fire' or the operation of a manual device such as a gong, whistle or air horn is not adequate to warn everyone when operated from any single point within the building, it is likely that an electrical fire warning system may be required. This will typically include the following:

- manual call points (break-glass call points) at storey exit and final exit doors;

- electronic sirens or bells; and

- a simple control and indicator panel.

Some premises (e.g. a community centre) may only require a few interconnected manual call points and sounders. Other premises (e.g. a multi-level pub) may require a more complex warning system involving staff action (see Section 2.3).

If your building has areas where a fire could develop undetected or where people work alone and might not see a fire (e.g. storerooms in a community centre), then it may be necessary to upgrade your fire-warning system to incorporate some form of automatic fire detection (e.g. in some circumstances, this could be a series of interconnected smoke alarms, if shown to be acceptable in your risk assessment).

If, for any reason, your system fails, you must still ensure that people in your premises can be warned and escape safely. A temporary arrangement, such as gongs, whistles or air horns, combined with suitable training, may be acceptable for a short period pending system repairs.

The fire warning sound levels should be loud enough to alert everyone, taking into account background noise, any sound systems should be muted (automatically or manually) when the fire alarm sounds. In areas with uncontrollable high background noise, or where people may be wearing hearing protectors, the audible warning should be supplemented, e.g. with visual alarms.

People with hearing difficulties

Where people have hearing difficulties, particularly those who are profoundly deaf, then simply hearing the fire warning is likely to be the major difficulty. If these persons are never alone while on the premises then this may not be a serious problem, as it would be reasonable for other occupants to let them know that the building should be evacuated. If a person with hearing difficulties is likely to be alone, then consider other means of raising the alarm. Among the most popular are visual beacons and vibrating devices or pagers that are linked to the existing fire alarm.

2.1 Manual call points

Manual call points, often known as 'break-glass' call points, enable a person who discovers a fire to immediately raise the alarm and warn other people in the premises of the danger.

People leaving a building because of a fire will normally leave by the way they entered. Consequently, manual call points are normally positioned at exits and storey exits that people may reasonably be expected to use in case of fire, not just those designated as fire exits. However, it is not necessary in every case to provide call points at every exit.

Manual call points should normally be positioned so that, after all fixtures and fittings are in place, no one should have to travel more than 45m to the nearest alarm point. This distance may need to be less if your premises cater for people of limited mobility or there are particularly hazardous areas. They should be conspicuous (red), fitted at a height of about 1.4m (or less for premises with a significant number of wheelchair users), and not in an area likely to be obstructed.

Conventionally sited call points that operate an immediate general alarm may not be desirable in some premises used for assembly.

To reduce the risk of malicious or accidental operation which may promote panic in the event of an actuation by a member of the public (e.g. in a pub), as as part of your management plan you should consider alternative arrangements such as:

- locating the majority of manual call points in staff areas;

- the guarding of manual call points; and

- a delayed alarm for investigation purposes before any general alarm is given.

Areas that are regularly checked and patrolled may have less need for a manual call point. For these approaches to be effective, robust management procedures should be in place around a staff alarm system (see Section 2.3 below).

2.2 Automatic fire detection

Automatic fire detection may be needed for a number of reasons. These can include:

- if you have areas where people are isolated or remote and could become trapped by a fire because they are unaware of its development, such as lone workers;

- if you have areas where a fire can develop unobserved (e.g. storerooms);

- as a compensating feature, e.g. for inadequate structural fire protection, in dead-ends or where there are extended travel distances; and

- where smoke control and ventilation systems are controlled by the automatic fire-detection system.

If you have an automatic fire detection system, the system should:

- be designed to accommodate the emergency evacuation procedure;

- give an automatic indication of the fire warning and its location;

- be maintained and tested by a competent person; and

- communicate with a central control room (if you have one).

New automatic fire detection systems should be designed and installed by a competent person. Further guidance is given in BS 5839-1[16] or a more recent standard where applicable.

2.3 Staff alarms

In some premises the actuation of the fire warning system should trigger the immediate and total evacuation of the building. However, in some premises this may not be necessary as alternative arrangements may be in place.

For example, the initial alert signal could be given only to certain staff, who then carry out pre-arranged actions to help others to evacuate more easily. This will require able, fully-trained staff to be available at all times, particularly to alert disabled people to give them the maximum time to escape. This approach should form part of your emergency plan.

Where the public address system is part of the fire warning system it should be connected to an auxiliary power source to ensure the continued use of the system in the event of fire or other emergency.

Whichever warning or detection systems are in place, however, if a fire occurs the fire and rescue service should always be called immediately.

2.4 Reducing false alarms

False alarms from automatic fire detection systems are a major problem and result in many unwanted calls to the fire and rescue service every year. Guidance on reducing false alarms has been published by ODPM/CFOA/BFPSA.[15]

If there are excessive false alarms in your premises, people may become complacent and not respond correctly to a warning of a real fire. In such circumstances, you may be failing to comply with fire safety law. All false alarms should be investigated to identify the cause of the problem and remedial action taken.

Further detailed guidance on reducing false alarms is available in BS 5839-1.[16]

2.5 Testing and maintenance

Your fire warning and/or detection system should be supervised by a named responsible person, who has been given enough authority and training to manage all aspects of the routine testing and scrutiny of the system.

The control and indicating equipment should be checked at least every 24 hours to ensure there are no specific faults. All types of fire warning systems should be tested once a week. For electrical systems a manual call point should be activated (using a different call point for each successive test), usually by inserting a dedicated test key (see Figure 18). This will check that the control equipment is capable of receiving a signal and in turn, activating the warning alarms. Manual call points may be numbered to ensure they are sequentially tested.

Figure 18: Using a test key

It is good practice to test the alarm at the same time each week, but additional tests may be required to ensure that staff or people present outside normal working hours are given the opportunity to hear the alarm.

Where systems are connected to a central monitoring station arrangements should be made prior to testing to avoid unwanted false alarms.

Six-monthly servicing and preventive maintenance should be carried out by a competent person with specialist knowledge of fire warning and automatic detection systems. This task is normally fulfilled by entering into a service contract with a specialist fire alarm company.

It is good practice to record all tests, false alarms and any maintenance carried out.

Further guidance on testing and maintenance of fire warning systems can be found in British Standard 5839-1.[16]

2.6 Guaranteed power supply

If your fire risk assessment concludes that an electrical fire-warning system is necessary, then the Health and Safety (Safety Signs and Signals) Regulations 1996[5] requires it to have a back-up power supply.

Whatever back-up system is used, it should normally be capable of operating the fire-warning and detection system for a minimum period of 24 hours and sounding the alarm signal in all areas for 30 minutes.

2.7 New and altered systems

Guidance on the design and installation of new systems and those undergoing substantial alterations is given in BS 5839-1.[16] If you are unsure that your existing system is adequate you will need to consult a competent person.

Section 3 Further guidance on firefighting equipment and facilities

All premises should be provided with appropriate firefighting eqipment. For the majority of premises, first aid firefighting equipment should be sufficient.

You have responsibility for the provision of appropriate firefighting equipment. It is also your responsibility to check that all firefighting equipment is in the correct position and in satisfactory order before the premises are used.

Appropriate staff should be trained in the use of all such equipment.

3.1 Portable firefighting equipment

Fire extinguishers provided should be appropriate to the specific risks found in your premises in accordance with Table 1. This table also shows the different classes of fire, according to what is burning.

At events such as exhibitions you must take care that temporary stands or exhibits do not obscure extinguishers or manual call points.

Number and type of extinguishers

Typically for the Class A fire risk, the provision of one water-based extinguisher for approximately every 200m² of floor space, with a minimum of two extinguishers per floor, will normally be adequate.

Where it is determined that there are additionally other classes of fire risk, the appropriate type, number and size of extinguisher should be provided. Further information is available in BS 5306-8.[18]

Table 1: Class of fire

Class of fire	Description
Class A	Fires involving solid materials such as wood, paper or textiles.
Class B	Fires involving flammable liquids such as petrol, diesel or oils.
Class C	Fires involving gases.
Class D	Fires involving metals.
Class F	Fires involving cooking oils such as deep-fat fryers.

Note: If there is a possibility of a fire in your premises involving material in the shaded boxes then you should seek advice from a competent person.

Where the fire risk is not confined to a particular location, e.g. Class A fires, the fire extinguishers should be positioned on escape routes, close to the exit from the room or floor, or the final exit from the building. Similarly, where the particular fire risk is specifically located, e.g. flammable liquids, the appropriate fire extinguisher should be near to the hazard, so located that they can be safely used. They should be placed on a dedicated stand or hung on a wall at a convenient height so that employees can easily lift them off (at about 1m for larger extinguishers, 1.5m for smaller ones, to the level of the handle). Ideally no one should have to travel more than 30m to reach a fire extinguisher. If there is a risk of malicious use you may need to use alternative, and more secure, locations.

Consider the implications of the Manual Handling Operations Regulations 1992[17] when selecting and siting firefighting equipment.

In self-contained small premises, multi-purpose extinguishers which can cover a range of risks may be appropriate. Depending on the outcome of your fire risk assessment, it may be possible to reduce this to one extinguisher in very small premises with a floor space of less than 90m².

Extinguishers manufactured to current standards (BS EN 3-7)[86] are predominately red but may have a colour-coded area, sited above or within the instructions, denoting the type of extinguisher. Most older extinguishers, manufactured to previous standards, have bodies painted entirely in a single colour which denotes the type of extinguisher. These older extinguishers remain acceptable until they are no longer serviceable. However, it is good practice to ensure that old and new style extinguishers are not mixed on the same floor of a building.

The following paragraphs describe the different types of extinguisher. The colour referred to is the colour of the extinguisher or the colour-coded area.

Water extinguishers (red)

This type of extinguisher can only be used on Class A fires. They allow the user to direct water onto a fire from a considerable distance. A 9-litre water extinguisher can be quite heavy and some water extinguishers with additives can achieve the same rating, although they are smaller and therefore considerably lighter. This type of extinguisher is not suitable for use on live electrical equipment.

Water extinguishers with additives (red)

This type of extinguisher is suitable for Class A fires. They can also be suitable for use on Class B fires and where appropriate, this will be indicated on the extinguisher. They are generally more efficient than conventional water extinguishers.

Foam extinguishers (cream)

This type of extinguisher can be used on Class A or B fires and is particularly suited to extinguishing liquid fires such as petrol and diesel. They should not be used on free-flowing liquid fires unless the operator has been specially trained, as these have the potential to rapidly spread the fire to adjacent material. This type of extinguisher is not suitable for deep-fat fryers or chip pans.

Powder extinguishers (blue)

This type of extinguisher can be used on most classes of fire and achieve a good 'knock down' of the fire. They can be used on fires involving electrical equipment but will almost certainly render that equipment useless. Because they do not cool the fire appreciably it can re-ignite. Powder extinguishers can create a loss of visibility and may affect people who have breathing problems and are not generally suitable for confined spaces.

Carbon dioxide extinguishers (black)

This type of extinguisher is particularly suitable for fires involving electrical equipment as they will extinguish a fire without causing any further damage (except in the case of some electronic equipment, e.g. computers). As with all fires involving electrical equipment, the power should be disconnected if possible.

Class 'F' extinguishers

This type of extinguisher is particularly suitable for commercial catering establishments with deep-fat fryers.

Selection, installation and maintenance of portable fire extinguishers

All portable fire extinguishers will require periodic inspection, maintenance and testing. Depending on local conditions such as the likelihood of vandalism or the environment where extinguishers are located, carry out brief checks to ensure that they remain serviceable. In normal conditions a monthly check should be enough. Maintenance by a competent person should be carried out annually.

New fire extinguishers should comply with BS EN 3-7.[86] Guidance on the selection and installation of fire extinguishers is given in BS 5306-8,[18] for maintenance in, BS 5306-3,[19] and for colour coding in BS 7863.[20]

Fire blankets

Fire blankets should be located in the vicinity of the fire hazard they are to be used on, but in a position that can be safely accessed in the event of a fire. They are classified as either light duty or heavy duty. Light-duty fire blankets are suitable for dealing with small fires in containers of cooking oils or fats and fires involving clothing. Heavy-duty fire blankets are for industrial use where there is a need for the blankets to resist penetration by molten materials.

3.2 Fixed firefighting installations

If your premises forms part of a larger building there may be fixed firefighting systems and other facilities which are installed within the structure of the building. These may include the following.

Hose reels

Permanent hose reels installed in accordance with the relevant British Standard (see BS EN 671-3[21]) provide an effective firefighting facility. They may offer an alternative, or be in addition to, portable firefighting equipment. A concern is that untrained people will stay and fight a fire when escape is the safest option. Where hose reels are installed, and your fire risk assessment expects relevant staff to use them in the initial stages of a fire, they should receive appropriate training.

Sprinkler systems

Sprinkler systems can be very effective in controlling fires. They can be designed to protect life and/or property and may be regarded as a cost-effective solution for reducing the risks created by fire. Where installed, a sprinkler system is usually part of a package of fire precautions in a building.

Sprinkler protection could give additional benefits, such as a reduction in insurance premiums, a reduction in the amount of portable firefighting equipment necessary, and the relaxation of restrictions in the design of buildings.

Guidance on the design and installation of new sprinkler systems and the maintenance of all systems is given in the Loss Prevention Council (LPC) rules,[89] BS EN 12845[22] and BS 5306-2[69] and should only be carried out by a competent person. Routine maintenance by on-site personnel may include checking of pressure guages, alarm systems, water supplies, any anti-freezing devices and automatic booster pump(s).

A competent maintenance contractor should provide guidance on what records need to be completed.

Other fixed installations

There are a number of other fixed installations including water mist, gaseous, deluge and fixed powder systems. If your premises have a fixed firefighting system that you are unfamiliar with, then seek advice. Where a fixed firefighting systems forms an integral part of your safety strategy, it should be maintained in accordance with the relevant standard by a competent person.

3.3 Other facilities (including those for firefighters)

Building Regulations and other Acts, including local Acts, may have required firefighting equipment and other facilities to be provided for the safety of people in the building and to help firefighters. Fire safety law places a duty on you to maintain such facilities in good working order at all times. These facilities could include the following.

Access for fire engines and firefighters

Buildings that have been constructed to modern building regulations or in accordance with certain local Acts will have been provided with facilities that allow fire engines to approach and park within a reasonable distance so that firefighters can use their equipment without too much difficulty.

These facilities may consist of access roads to the building, hard standing areas for fire engines and access into the building for firefighters. It is essential that where such facilities are provided they are properly maintained and available for use at all relevant times.

Where a building is used by a number of different occupants you will need to ensure co-operation between the various 'responsible people' to maintain fire and rescue service access. In exceptional cases, where access is persistently obstructed, you may need to make additional arrangements.

See Approved Document B to the Building Regulations[24] for more information.

Firefighting shafts and lifts

Firefighting shafts are provided in larger buildings to help firefighters reach floors farther away from the building's access point. They enable firefighting operations to start quickly and in comparative safety by providing a safe route from the point of entry to the floor where the fire has occurred.

Where a firefighting shaft is provided it should be maintained by a competent person.

Smoke control systems

These are complex systems that are provided for life safety of occupants, assistance to firefighters and property protection by clearing hot smoke and gases from the building. If you have one of these systems provided in your premises you should ensure that you understand how it operates and that it is maintained in full working order. If your system is part of a larger system then you should liaise with other occupiers and building managers.

The smoke control system should be maintained by a competent person who is familiar with the fire engineering performance specifications of that specific system.

Dry and wet rising fire mains

The rising fire main is an important facility for the fire and rescue service in taller buildings. It consists of an inlet box where firefighters can connect their hoses, a pipe running up or through the building, outlet valves on each floor level and an air vent at the top.

Firefighters' switches

Safety switches are normally provided to isolate high-voltage luminous signs or to cut off electrical power. In the case of existing installations, if they have been provided in accordance with previous legislation (e.g. the Local Government (Miscellaneous Provisions) Act 1982[25]), then it is likely that they will comply with the Order. If this is not the case, then you may need to consult the enforcing authority regarding the suitability of its location and marking.

You should consult with the building management (e.g. a health and safety manager of a complex) to establish if you have any of these facilities and, if you have, you will need to co-operate with management to ensure that such facilities are maintained by a competent person.

Section 4 Further guidance on escape routes

Introduction

This section provides further guidance on the general principles that apply to escape routes and provides examples of typical escape route solutions for a range of building layouts. The guidance is based on premises of normal risk so if your premises (or part of your premises) are higher (or lower) risk you should adopt the solution accordingly.

You are not obliged to adopt any particular solution for escape routes in this section if you prefer to meet the relevant requirement in some other way. If you decide to adopt some alternative arrangement it will need to achieve at least an equivalent level of fire safety.

Refer to the glossary (Appendix D) for the definitions of any terms you may not be familiar with.

Levels of risk

In order to apply the guidance in this section, you need to understand that in any fire situation, the time that people have to escape before they could become affected by the fire is limited. Providing them with sufficient time usually means that as well as having an appropriate way of detecting and giving warning in case of fire, the distance that people have to travel to make their escape to a place of reasonable or total safety must be restricted.

The travel distances which are usually appropriate for this purpose (and are suggested later in this section) vary according to the level of risk in the premises (or part of them). To check your escape routes you will need to form a judgement about the level of risk that people may be at after you have taken other risk reduction (preventative and protective) measures.

In premises where there is a likelihood of a fire starting and spreading quickly (or a fire could start and grow without being quickly detected and a warning given) and affect the escape routes before people are able to use them then the risk should normally be regarded as 'higher'. Such premises could include those where significant quantities of flammable materials are used or stored; ready sources of ignition are present, e.g. heat producing machinery and processes; premises where significant numbers of the people present are likely to move slowly or be unable to move without assistance; and premises where the construction provides hidden voids or flues through which a fire could quickly spread.

In premises where there is a low occupancy level and all the occupants are able bodied and capable of using the means of escape without assistance; very little chance of a fire; few if any highly combustible or flammable materials or other fuels for a fire; fire cannot spread quickly; and will be quickly detected so people will quickly know that a fire has occurred and can make their escape, then the risk can usually be regarded as 'lower'.

In most cases however, the risk will usually be 'normal'.

The travel distances suggested are not hard and fast rules and should be applied with a degree of flexibility according to the circumstances. For example, in premises where the risk might otherwise be considered 'normal' but where there are a significant number of people who move slowly or may need assistance to evacuate, it would usually be appropriate to consider this a 'higher' risk. However, where other measures are in place to mitigate this, such as the availability of extra assistance and this has been planned for in your emergency plan, it may be that the risk level can be regarded as 'normal to higher'.

Equally, in premises where the risk category would otherwise be 'lower' but for the fact that a small number of occupants may move slowly or need assistance, it may be appropriate to categorise the risk as 'normal' in these circumstances.

If you are not sure about the level of risk that remains in your premises, you should seek advice from a competent person.

4.1 General principles

Suitability of escape routes

You should ensure that your escape routes are:

- suitable;

- easily, safely and immediately usable at all times;

- adequate for the number of people likely to use them;

- free from any obstructions, slip or trip hazards;

- well lit by normal or emergency escape lighting; and

- available for access by the emergency services.

In multi-occupied premises, escape routes should normally be independent of other occupiers, i.e. people should not have to go through another occupier's premises as the route may be secured or obstructed. Where this is not possible, then robust legal agreements should be in place to ensure their availability at all times.

All doors on escape routes should open in the direction of escape, and ideally be fitted with a safety vision panel. This is particularly important if more than 60 people are expected to use them at any one time or they provide an exit from an area of high fire risk.

At least two exits should be provided if a room/area is to be occupied by more than 60 persons. This number of 60 can be varied in proportion to the risk, for a lower risk there can be a slight increase, for a higher risk, lower numbers of persons should be allowed.

Movement of persons up or down a group of not less than three steps will be so obvious to those following that they will be prepared for the change in level, but movement up or down one step is not so readily observed and may easily lead to a fall. Wherever practicable, differences of level in corridors, passages and lobbies should be overcome by the provision of inclines or ramps of gradients not exceeding 1 in 12 or steps not having less than three risers in any flight. Corridors and passages should be level for a distance of 1.5 metres in each direction from any steps.

Any mirrors situated in escape routes should be sited so that persons escaping from a fire will not be thrown into confusion by any reflected image of the route they are using, or be misled as to the direction they should take to reach fire exits.

In historic buildings (e.g. churches) you may have difficulties achieving these recommendations and your risk assessment should address any implications, for example through your management system. Further information on listed and historic buildings is given in Appendix C.

Seating and gangways

The type of seating arrangements adopted will vary with the use to which the premises are put. Premises should only be used for closely-seated audiences if your risk assessment shows that it is safe to do so.

Where premises are licensed, there may be additional conditions in the licence concerning how seating may be set out.

Seating and gangways in a hall or assembly space should be so arranged to allow free and ready access direct to the exits.

Audiences seated in rows will first have to make their way to the end of the row before being able to use the escape routes provided. Seating and gangways in an auditorium should therefore be so arranged as to allow free and ready access direct to the exits.

In fixed seats there should be a clear space of at least 305mm between the back of one seat to the front of the seat behind it (or the nearmost point of the seat behind it, for automatic tip-up seats (see Figure 19). Gangways should be adequate for the number of seats served and at least 1.05m wide. There should be no projections which diminish these widths.

In general, no seat should be more than seven seats away from a gangway. If temporary seating is provided, these should be secured in lengths of not fewer than four seats (and not more than 12). Each length should be fixed to the floor.

Detailed information of seating layout is given in BS 5588-6.[68]

Standing and sitting in gangways, or in front of any exit, should not be permitted. This includes adults standing by children.

If booster seats are provided for small children, then you need to consider the implication on means of escape.

Figure 19: Clear space between seating

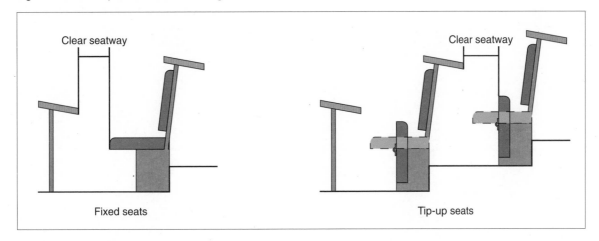

Clear seatway

Fixed seats

Clear seatway

Tip-up seats

Fire-resisting construction

The type and age of construction are crucial factors to consider when assessing the adequacy of the existing escape routes. To ensure the safety of people it may be necessary to protect escape routes from the effects of a fire. In older premises (see Appendix C for more information on historical properties) it is possible that the type of construction and materials used may not perform to current fire standards. Also changes of occupancy and refurbishment may have led to:

- cavities and voids being created, allowing the potential for a fire to spread unseen;

- doors and hardware worn by age and movement being less likely to limit the spread of smoke;

- damaged or lack of cavity barriers in modular construction; and

- breaches in fire compartment walls, floors and ceilings created by the installation of new services, e.g. computer cabling.

Where an escape route needs to be separated from the rest of the premises by fire-resisting construction, e.g. a dead-end corridor or protected stairway (Figure 28), then you should ensure the following:

- Doors (including access hatches to cupboards, ducts and vertical shafts linking floors), walls, floors and ceilings protecting escape routes should be capable of resisting the passage of smoke and fire for long enough so that people can escape from the building.

- Where suspended or false ceilings are provided, the fire resistance should extend up to the floor slab level above. For means of escape purposes a 30-minute fire-resisting rating is usually enough.

- Cavity barriers, fire stopping and dampers in ducts are appropriately installed.

If there is any doubt about the nature of the construction of your premises, ask for advice from a competent person.

Number of people using the premises

As your escape routes need to be adequate for the people likely to use them you will need to consider how many people, including employees and the public, may be present at any one time. Where premises have been subject to building regulations approval, the number and width of escape routes and exits will normally be enough for the anticipated number of people using the premises. In such buildings where the risk has changed or buildings were constructed before national building regulations, it will be necessary to confirm the provision.

For most premises, the maximum numbers of staff, visitors and contractors liable to be in the building at the same time will be known by the responsible person. For some premises (e.g. sports centres or churches) the responsible person will normally be aware of the maximum number of people liable to be present from a personal knowledge of use patterns. There will also be an appreciation of the use of the building by those with special needs such as disabled people.

You also need to be aware of any licensing requirements and capacity limits set by your license, although these may not be directly related to fire safety.

Where you have fixed seating then the maximum numbers of people in the hall or auditorium will be known from the number of seats provided. If you are providing temporary seating then this may be limited by the maximum numbers your escape routes can accommodate (see case study below). In un-seated areas, foyers or bars you may estimate the maximum numbers based on a value of 0.3 to 0.5m² per person.

If you propose to make changes to the use or layout of the building which may increase the number of people, you should check the design capacity by referring to guidance given in BS 5588 Part 6[68] and the Building Regulations Approved Document B[24] and consult the Building Control Body.

If the numbers of people entering the premises may be too great then procedures may be needed to control entry numbers, for example the use of tickets and clickers, or restrictions to members in private clubs.

In some situations you may need to control numbers within specific areas and/or prevent migration and overcrowding. Different events may require different responses, for example, ticketed events and events with free admission.

Case study

During the Christmas period, a public house became severely overcrowded due to a large number of office parties. The number of people became so large it would have been unlikely, in the event of a fire, for all members of the public to evacuate the premises in an acceptable time.

The number of people entering the premises should be controlled to avoid this overcrowding.

Mobility impairment
Effective management arrangements need to be put in place for those that need help to escape.

Consider the following points:

- A refuge is a place of reasonable safety in which disabled people can wait either for an evacuation lift or for assistance up or down stairs (see Figure 20). Disabled people should not be left alone in a refuge area whilst waiting for assistance to evacuate the building. Depending on the design and fire resistance of other elements, a refuge could be a lobby, corridor, part of a public area or stairway, or an open space such as a balcony or similar place which is sufficiently

protected (or remote) from any fire risk and provided with its own means of escape and a means of communication.

- Where refuges are provided, they should be enclosed in a fire-resisting structure which creates a protected escape route which leads directly to a place of total safety and should only be used in conjunction with effective management rescue arrangements. Your fire safety strategy should not rely on the fire and rescue service rescuing people waiting in these refuges.

- If firefighting lifts (provided in high buildings as firefighting access) are to be used for evacuation, this should be co-ordinated with the fire and rescue service as part of the pre-planned evacuation procedures.

- Normal lifts may be considered suitable for fire evacuation purposes, subject to an adequate fire risk assessment and development of a suitable fire safety strategy by a competent person.

- Since evacuation lifts can fail, a disabled person, having reached a refuge, should also be able to gain access to a stairway (should conditions in the refuge become untenable). An evacuation lift with its associated refuge should therefore be located adjacent to a protected stairway.

- Enough escape routes should always be available for use by disabled people. This does not mean that every exit will need to be adapted. Staff should be aware of routes suitable for disabled people so that they can direct and help people accordingly.

- Stairways used for the emergency evacuation of disabled people should comply with the requirements for internal stairs in the building regulations. Specialist evacuation chairs or other equipment may be necessary to negotiate stairs.

- Plans should allow for the careful carrying of disabled people down stairs without their wheelchairs, should the wheelchair be too large or heavy. You will need to take into account health and safety manual handling procedures in addition to the dignity and confidence of the disabled person.

- Stairlifts should not be used for emergency evacuation. Where installed in a stairway used for emergency evacuation, no parts of the lift, such as its carriage rail, should be allowed to reduce the effective width of the stairway or any other part of an emergency evacuation route.

- Where ramps are necessary for the emergency evacuation of people in wheelchairs they should be as gentle as possible. Ramps should be constructed in accordance with Approved Document M.[71]

Further guidance is available in BS 5588-8[72] and BS 5588-12.[51]

Figure 20: An example of a refuge

Widths and capacity of escape routes and stairways

Once you have established the maximum number of people likely to be in any part of the premises, the next step is to establish that the capacity of the escape routes is adequate for people to escape safely and in sufficient time to ensure their safety in case of fire.

The capacity of a route is determined by a number of factors including the width of the route, the time available for escape and the ability of the persons using them.

The effective usable width of an escape route is the narrowest point, normally a door or other restriction such as narrowing of a corridor due to fixtures and fittings. The capacity of an escape route is measured by the number of persons per minute that can pass through it so, to establish the capacity of the route, it is first necessary to measure the width of the route at the narrowest point. The effective width of a doorway is the clear unobstructed width through the doorway when the door is open at right angles to the frame. The effective width at any other point is the narrowest clear unobstructed width through which people can pass.

The time available for escape depends on several factors including the distance that has to be travelled to reach a place of safety and the risk rating of the premises. Established reasonable escape times are 2 minutes for higher risk premises, 2.5 minutes for normal risk premises and 3 minutes for lower risk premises. For calculation purposes these times are allowed for in the travel distances suggested in Table 2. Guidance on establishing the risk rating of your premises is given earlier in this section.

The following guide can be used to determine the general capacities of escape routes:

A width of at least 750mm can accommodate up to:

- 80 people in higher risk premises;

- 100 people in normal risk premises; or

- 120 people in lower risk premises.

A width of at least 1,050mm can accommodate up to:

- 160 people in higher risk premises;

- 200 people in normal risk premises; or

- 240 people in lower risk premises.

An additional 75mm should be allowed for each additional 15 persons (or part of 15).

The minimum width of an escape route should ideally be 1,050mm but in any case not less than 750mm (unless it is for use by less than five people in part of your premises) and, where wheelchair users are likely to use it, not less than 900mm.

The aggregate width of all the escape routes should be not less than that required to accommodate the maximum numbers of people likely to use them.

When calculating the overall available escape route capacity for premises that have more than one way out, you should normally assume that the widest is not available because it has been compromised by fire. If doors or other exits leading to escape routes are too close to one another you should consider whether the fire could affect both at the same time. If that is the case, it may be necessary to discount them both from your calculation.

As a general rule stairways should be at least 1,050mm wide and in any case not less than the width of the escape routes that lead to them. In all cases the aggregate capacity of the stairways should be sufficient for the number of people likely to have to use them in case of fire.

Stairways wider than 2,100mm should normally be divided into sections, each separated from the adjacent section by a handrail, so that each section measured between the handrails is not less than 1,050mm wide.

Mirrors
Any mirrors situated on escape routes should be so sited that persons escaping from fire will not be confused by any reflected image of the route they are using, or be misled as to the direction they should take to reach fire exits.

Travel distance
Having established the number and location of people and the exit capacity required to evacuate them safely, you now need to confirm that the number and location of existing exits is adequate. This is normally determined by the distance people have to travel to reach them.

Table 2 gives guidance on travel distances. It should be understood, however, that these distances are flexible and may be increased or decreased depending upon the level of risk after you have put in place the appropriate fire-prevention measures (Part 1, Step 3.3).

In new buildings which have been designed and constructed in accordance with modern building standards the travel distances will already have been calculated. Once you have completed your fire risk assessment you need to confirm that those distances are still relevant.

When assessing travel distances you need to consider the distance to be travelled by people when escaping, allowing for walking around furniture or display stands, etc. The distance should be measured from all parts of the premises to the nearest place of reasonable safety which is:

- a protected stairway enclosure (a storey exit);

- a separate fire compartment from which there is a final exit to a place of total safety; or

- the nearest available final exit.

Table 2: Suggested travel distances (not for marquees)

Escape routes	Suggested range of travel distance: areas with seating in rows	Suggested range of travel distance: other areas
Where more than one route is provided	20m in higher fire-risk area[1] 32m in normal fire-risk area 45m in lower fire-risk area[2]	25m in higher fire-risk area[1] 45m in normal fire-risk area 60m in lower fire-risk area[2]
Where only a single escape route is provided	10m in higher fire-risk area[1] 15m in normal fire-risk area 18m in lower fire-risk area[2]	12m in higher fire-risk area[1] 18m in normal fire-risk area 25m in lower fire-risk area[2]

Note 1:
Where there are small high risk areas this travel distance should apply. Where the risk assessment indicates that the whole building is high risk, seek advice from a competent person.

Note 2:
The travel distance for low risk premises should only be applied in exceptional cases in the very lowest risk premises where densities are low, occupants are familiar with the premises, there is excellent visual awareness, and very limited combustibles.

The travel distances given in Table 2 are based on those recommended in Approved Document B of the Building Regulations[24] and are intended to complement the other fire safety recommendations in Approved Document B. Your current escape route travel distances may be different from these since they may be based on recommendations made in alternative guidance.

Where your route leads to more than one final exit, but only allows initial travel in a single direction (e.g. from a room or dead end, see Figures 25 and 26), then this initial travel distance should be limited to that for a 'single escape route' in Table 2. However, your total travel distance should not exceed that for 'more than one escape route'.

For marquees, the travel distance from any part of the structure having more than one exit should be 24m, the initial 6.5m of which may be in a single direction of travel. Where there is only one exit, the travel distance should not exceed 6.5m.

Measuring travel distance
The figures that follow are schematic only and are intended to represent part of a larger building.

The route taken through a room or space will be determined by the layout of the contents, e.g. seating (Figure 21). It is good practice to ensure routes to the exits are kept as direct and short as possible. In a small room there may be only one exit but in a larger room or area there may be many exits.

Figure 21: Measuring travel distance

In some cases, where the contents are moved around or the space is liable to frequent change, e.g. in a storage area or where racking is movable, you should ensure that the exits, or the routes to them, do not become blocked or the length of the route is not significantly extended.

Inner rooms

Where the only way out of a room is through another room (Figure 22), an unnoticed fire in the outer room could trap people in the inner room. This layout should be avoided where possible. If, however, this cannot be achieved then adequate warning of a fire should be provided by **any one** of the following means:

- a vision panel between the two rooms providing adequate vision to give an indication of the conditions in the outer room and the means of escape;

- a large enough gap between the dividing wall and the ceiling, e.g. 500mm, so that smoke will be seen; or

- an automatic smoke detector in the outer room that will sound a warning in the inner room.

In addition, the following points should also be considered:

- Restrict the number of people using an inner room to 60.

- Access rooms should be under the control of the same person as the inner room.

- The travel distance from any point in the inner room to the exit from the access room should be restricted to escape in one direction only (see Table 2), unless there are alternative exits from the access room.

- No one should have to pass through more than one access room while making their escape.

- The outer room should not be an area of high fire risk.

Figure 22: Inner rooms

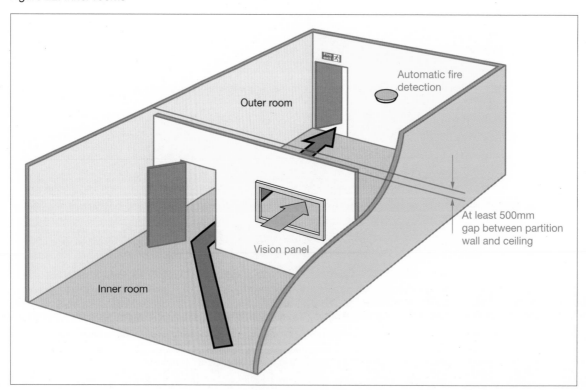

Alternative exits

Where alternative exits from a space or room are necessary they should wherever possible be located at least 45° apart (see Figure 23) unless the routes to them are separated by fire-resisting construction (see Figure 24). If in doubt consult a competent person.

Figure 23: Alternative exits

Figure 24: Alternative exits (seperated by fire-resisting construction)

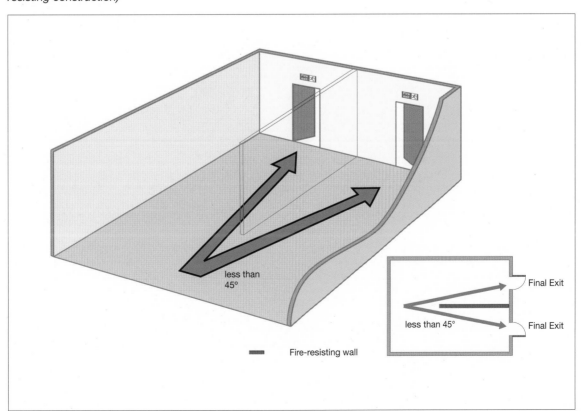

Measuring travel distances for initial dead-end travel

Where the initial direction of travel in an open area (see Figure 25) is in one direction only or within an inner room (see Figure 26), the travel distance (A–B) should be limited to that for a 'single escape route' in Table 2.

Any alternative exits should be positioned to ensure that a fire will not compromise both exits. The maximum total travel distance recommended in Table 2 should apply to the nearest exit (Figure 26, distance A–C). However, since you have two exits, your total travel distance should not exceed that for 'more than one escape route' in Table 2.

Figure 25: Measuring travel distance from initial dead end (open plan)

A–B	Travel in single direction only
A–C	Total travel distance (to nearest exit)

Figure 26: Measuring travel distance from initial dead end (inner room)

A–B   Travel in single direction only

A–C   Total travel distance (to nearest exit)

Note: Additional measures are necessary to give warning of fire for inner rooms (see Figure 22).

Escape routes with dead-end conditions

If your premises has escape routes from which escape can be made in one direction only (a dead end), then an undetected fire in that area could affect people trying to escape. To overcome this problem, limit the travel distance (see Table 2) and use one of the following solutions:

- Fit an automatic fire detection and warning system in those areas where a fire could present a risk to the escape route (see Figure 27).

- Protect the escape route with fire-resisting construction to allow people to escape safely past a room in which there is a fire (see Figure 28).

- Provide an alternative exit (see Figure 29).

Alternative approaches may be acceptable, although expert advice may be necessary.

Figure 27: Dead-end condition with automatic fire detection

A–B	Travel in single direction only
A–C	Total travel distance
⬭	Automatic fire detection

Figure 28: Dead-end condition with fire-resisting construction

A–B	Travel in single direction only
A–C	Total travel distance
SC	Self-closing
▬	Fire-resisting wall
⌐	Fire door

Figure 29: Dead-end condition provided with an alternative exit

A–B   Travel in single direction only
A–C   Total travel distance

Basements, escape and protection

In all buildings with basements (other than very small basements), stairways serving upper floors should preferably not extend to the basement and in any case should not do so where they are the only stairway serving the upper floors. Any stairway that extends from the basement to the upper floors should be separated at basement level by a fire-resisting lobby or corridor between the basement and the stairway. All basements used by more than 60 people or where there are no exits directly to a place of total safety, should have at least two protected escape stairways.

In high risk premises there should be an alternative stairway from the basement to ground level, unless there is a suitable alternative route to the final exit.

Wherever possible all stairways to basements should be entered at ground level from the open air, and should be positioned so that smoke from any fire in the basement would not obstruct any exit serving the other floors of the building.

Where any stairway links a basement with the ground floor, the basement should be separated from the ground floor, preferably by two 30-minute fire doors one at basement and one at ground floor level (see Figure 30).

As a minimum any floor over a basement should provide 60 minutes fire resistance. For smaller premises 30 minutes may be acceptable. Where this is impractical, and as long as no smoke can get through the floor, automatic smoke detection linked to a fire-alarm system which is audible throughout the premises could, as an alternative, be provided in the basement area. If in doubt, contact a competent person for more detailed advice.

Figure 30: Basement protection

Ground floor

Basement

SC

SC

	Fire-resisting wall
⌐|	Fire door
SC	Self-closing

Subdivision of corridors

If your premises has corridors more than 30m long, then generally these corridors should be subdivided near the centre of the corridor with fire doors and, where necessary, fire-resisting construction to limit the spread of fire and smoke and to protect escape routes if there is a fire.

Where a corridor serves two exits from a floor, generally these corridors should be subdivided with fire doors to separate the two exits (see Figure 31).

Doors that are provided solely for the purpose of restricting the travel of smoke need not be fire doors, but will be suitable as long as they are of substantial construction, are capable of resisting the passage of smoke, and are self-closing. Smoke should not be able to bypass these doors, e.g. above a false ceiling, or via alternative doors from a room, or adjoining rooms, opening on either side of the subdivision.

Generally, false ceilings should be provided with barriers or smoke stopping over any fire doors. Where the false ceiling forms part of the fire-resisting construction this may not be necessary.

If you have doubts about subdivision of corridors, ask advice from a competent person.

Figure 31: Subdivision of corridor between two stairways or exits

Fire door

Fire-resisting wall

SC   Self-closing

Upper storey

Stairway enclosures

Stairways, if unprotected from fire, can rapidly become affected by heat and smoke, cutting off the escape route and allowing fire spread to other floors. However, if adequately protected, escape stairways can be regarded as places of reasonable safety to enable people to escape to a place of total safety.

In most premises designed and built to building regulations and served by more than one stairway, it is probable that these stairways will be protected by fire-resisting construction and will lead to final exit. If any floor has an occupancy of over 60, each storey should have at least two exits, i.e. protected routes. The figure of 60 can be varied in proportion to the risk, lower risk slight increase, higher risk lower numbers of persons.

It is possible that you may have some stairways which have no fire protection to them. In this case they are not designed for escape and are normally known as accommodation stairways (see accommodation stairways on page 77).

If you have a protected stairway(s) then it is essential that you maintain that level of fire protection.

The benefit of protecting stairways from the effects of fire allows you to measure your travel distance from the furthest point on the relevant floor to the nearest storey exit rather than the final exit of the building.

If you do not have a protected stairway, depending on the outcome of your fire risk assessment, it may be that you can achieve an equivalent level of safety by other means. However, before doing so you should seek advice from a competent person.

If the building you occupy has floors which are occupied by different organisations to your own, you need to consider, as part of your fire risk assessment, the possibility that a fire may occur in another part of the building over which you may have no control and which may affect the protected stairway if allowed to develop unchecked. If your fire risk assessment shows that this may be the case and people using any floor would be unaware of a developing fire, then additional fire-protection measures may be required, e.g. an automatic fire-detection and warning system. If this is required you will need to consult and co-operate with other occupiers and building managers.

You may find that stairways in your building are provided with protected lobbies or corridors at each floor level (see Figure 32). Although these are not generally necessary for means of escape

Figure 32: Examples of a stairway with protected lobby/corridor approach

Protected corridor

Fire door
Fire-resisting wall
SC   Self-closing

Protected lobby

Fire door
Fire-resisting wall
SC   Self-closing

in multi-stairway buildings of less than 18m high, they may have been provided for other reasons (e.g. firefighting access). In all cases protected corridors, lobbies and stairways must be kept clear of combustibles and obstructions.

Ideally stairway enclosures should lead directly to a final exit. If your premises has only one stairway from the upper floor(s) which does

not lead directly to a final exit, adopt one of the following arrangements:

*   provide a protected route from the foot of the stairway enclosure leading to a final exit (see Figure 33); or

*   provide two exits from the stairway, each giving access to a final exit via routes which are separated from each other by fire-resisting construction (see Figure 34).

Figure 33: Examples of a protected route from a stairway to a final exit

Figure 34: Examples of two escape routes from a stairway to final exits

Separation of protected stairways

Where there are two or more protected stairways, the routes to final exits should be separated by fire-resisting construction so that fire cannot affect more than one escape route at the same time (see Figure 35).

Figure 35: Separation of protected stairways

Creating a stairway bypass route

No one should have to pass through a protected stairway to reach another stairway. Options to avoid this include:

- using intercommunicating doors between rooms adjacent to the stairway; such doors must be available at all times when the building is occupied (see Figure 36);

- using balconies and other features to bypass the stairway; or

- as long as there is enough space, create a bypass corridor around the stairway enclosure.

Figure 36: A stairway bypass route

Fire door

Fire-resisting wall

SC Self-closing

Reception areas

Reception or enquiry areas should only be located in protected stairways where the stairway is not the only one serving the upper floors, the reception area is small (less than 10m²) and is of low fire risk.

Accommodation stairways

If you have stairways that are used for general communication and movement of people in the premises, and they are not designated as fire escape stairs then these are called 'accommodation stairways'. They may not require fire separation from the remainder of the floor as long as they do not pass through a compartment floor, or people do not have to pass the head of such a stairway in order to access a means of escape stairway. However, experience shows that many people will continue to use these as an escape route.

Accommodation stairways should not normally form an integral part of the calculated escape route; however, where your fire risk assessment indicates that it is safe to do so, then you may consider them for that purpose. In these cases it may be necessary to seek advice from a competent person to verify this.

External stairways

To be considered a viable escape route, an external stairway should normally be protected from the effects of a fire along its full length. This means that any door, window (other than toilet windows) and walls within 1.8m horizontally and 9m vertically below any part of the stairway should be fire-resisting. Windows should be fixed shut and doors self-closing (see Figure 37).

Consider protecting the external stairway from the weather as the treads may become slippery, e.g. due to algae, moss or ice. If this is not possible, you must ensure that the stairway is regularly maintained. Consider fixing non-slip material to the treads.

External stairways should not normally be used for members of the public, particularly where large numbers are likely to use them.

Figure 37: Protection to an external stairway

1.8 m

1.8 m

Defined zone for fire-resisting walls, doors and windows on an external stairway

Spiral and helical stairways

Spiral and helical stairways are usually acceptable only in exceptional situations, e.g. for a maximum of 50 people who are not members of the public. The stairway should not be more than 9m in total height and not less than 1.5m in diameter with adequate headroom. A handrail should be continuous throughout the full length of the stairway.

However, spiral and helical stairways may be used as means of escape by more than 50 staff and may be used by the public if the stairways have been designed for that purpose. Further guidance is given in BS 5395-2,[26] including about type E (public) stairs under that standard.

Lifts

Due to the danger of the power supplies to a lift being affected by a fire, lifts not specifically designed as 'firefighting' or 'evacuation' lifts are not normally considered acceptable as a means of escape. However, where a lift and stairway for a means of escape are incorporated in a fire-resisting shaft which has a final exit from it at the access level and the lift has a separate electrical supply to that of the remainder of the building, then that lift, subject to an agreed fire risk assessment, may be acceptable as a means of escape in case of fire.

Lifts are housed in vertical shafts that interconnect floors and compartments, therefore precautions have to be taken to protect people from the risk of fire and smoke spreading from floor to floor via the lift shaft. Such precautions may include:

- separating the lift from the remainder of the storey using fire-resisting construction and access via a fire door;

- ensuring the lift shaft is situated in a protected enclosure which may also be a stairway enclosure; and

- providing ventilation of at least 0.1m² at the top of each lift-well to exhaust any smoke.

Upward escape

Other than from a basement, people should not normally have to ascend more than one level to escape (e.g. from the top floor to use a roof exit). The upper level should not be an area of high fire risk.

Escalators and travelators

Escalators and travelators should not be used as part of the escape route unless they become stationary immediately on actuation of the alarm; they should then only be used as a means of escape if they satisfy the recommendations given for accommodation stairways.

Escalators not within stairway enclosures may need to be separated from the remainder of the building by fire-resisting construction and by fire doors.

To avoid a situation where persons are carried towards a fire it is important that arrangements should be made to stop escalators at the outset of an emergency.

Portable ladders and throw-out ladders

Throw-out ladders are not acceptable as a means of escape in a place of assembly. Portable ladders are only acceptable in exceptional circumstances where it will be reasonable for ladders of this kind to provide escape for one or two able-bodied people (e.g. from a high level plant room).

Fixed vertical or raking ladders

These are normally only suitable for use by a limited number of able-bodied staff where they form the only means of access to plant spaces, provided they are suitably guarded and the total descent does not exceed 9m without an intermediate landing. Where they are external they should be protected from the elements in the same way as external stairways.

Lowering lines and other self-rescue devices

These are not acceptable as a means of escape from a place of assembly.

Roof exits

It may be reasonable for the escape route to cross a roof. Where this is the case, additional precautions will normally be necessary:

- The roof should be flat and the route across it should be adequately defined and well-illuminated where necessary with normal electric and emergency escape lighting. The route should be non-slip and guarded with a protective barrier.

Figure 38: An escape route across a roof

- The escape route across the roof and its supporting structure should be constructed as a fire-resisting floor.

- Where there are no alternatives other than to use a roof exit, any doors, windows, roof lights and ducting within 3m of the escape route should be fire-resisting.

- The exit from the roof should be in, or lead to, a place of reasonable safety where people can quickly move to a place of total safety.

- Where an escape route passes through or across another person's property, you will need to have a robust legal agreement in place to allow its use at all times where people are on your premises.

- These routes should not normally be used by members of the public.

A typical escape route across a roof is illustrated in Figure 38.

External escape routes should receive routine inspection and maintenance to ensure they remain fit for use.

If your premises are part of a multi-occupied building, you will need to liaise with the owner or other persons responsible for the building. You will need to ensure that any legal agreements in place cover access for maintenance of the escape route.

Revolving doors, wicket doors, sliding doors and roller shutter

Revolving doors should not normally be considered as escape doors unless the leaves fold outward to form a clear opening upon pressure from within, or standard doors of the required exit width are provided next to the revolving door.

Ideally wicket doors or gates should have a minimum opening height of 1.5m. The bottom of the door should not be more than 250mm above the floor and the width should be preferably more than 500mm but not less than 450mm. Normally wicket doors will only be suitable for up to 15 members of staff; however, in areas of a higher fire risk, this should be reduced to a maximum of three.

Loading and goods delivery doors, shutters (roller, folding or sliding), up-and-over doors and similar openings are not normally suitable for use as a final exit. However, they may be suitable for escape from areas of normal risk by small numbers of staff as long as they are not likely to be obstructed and can be easily and immediately opened manually, even if normally power-operated, and the staff are familiar with the escape routes.

Sliding doors are not normally suitable on escape routes unless they are for the sole use of members of staff. Where provided, a notice with the words 'slide to open', with an arrow pointing in the direction of opening, should be permanently displayed at about eye level on the face of doors.

Fastenings on doors

You should ensure that any exit door or gate which is required to be kept open whilst the public are on the premises should be provided with appropriate fastenings so that the door or gate is kept open in such a way that it will not cause an obstruction. Appropriate fastenings for the public are push bars and push pads; handles requiring a turning action, for example, should not be used on doors for the public (see BS EN 1125[42] and BS EN 179[43]).

Doors used for means of escape should be kept unlocked at all times when people are in the premises and in no case should a door be so fastened that it cannot easily and immediately be opened from the inside without the use of a key. If the door has to be kept fastened while persons are in the building, the fastening should be by means of a panic latch or panic bar (or similar) so that the door can be readily opened by pressure applied by persons within.

For structural reasons some doors may not open in the direction of the exit. These, and all collapsible, sliding or roller gates, or shutters, should be locked (with a removable key) in the fully open position, at all times when the public are present. During such times, you should remove the key and keep it secure (e.g. in a manager's office).

If you have doors, gates, or shutters which need to be locked open, a notice or notices clearly visible on both sides of the door, gate or shutter (whether open or closed) should be prominently displayed. The notice should bear the words 'This door (gate, shutter, etc.) to be secured open when the premises are occupied'.

You should ensure that security fastenings such as chains, bars, padlocks, etc. are removed from all doors, gates, shutters, etc. on exit routes at all times the public are in the premises prior to the building being put to use.

No removable fastening should be replaced until the public have left the premises and sufficient exits should remain available for staff who may be present.

Exit doors and gates for events involving the public

Exit doors and gates should meet the following recommendations:

- All doors and gates should be capable of being immediately opened by those using that route in an emergency.

- All exit doors (except where there are less than 60 persons) and gates on an exit should open in the direction of escape so that people can escape safety in an emergency without obstruction.

- Sliding or roller-shutter gates and reversible turnstiles should not form any part of the emergency exit system.

- When open, no door should obstruct any gangway passage, stairway or landing.

Final exit doors and escape away from the premises

Good escape routes to a final exit will be of little benefit if the occupants are not able to get out of the building and quickly disperse from the area to a place of total safety. It is also important to consider where people will go once they have evacuated from the premises.

The matters that you should consider include the following:

- Final exit doors should be quickly and easily openable without a key or code in the event of a fire. Where possible, there should be only one fastening. See Appendix B3 for more information on security fastenings.

- Final exit doors should not lead people into an enclosed area from which there is no further escape.

- Where a final exit discharges into an enclosed area, further access to a place of total safety should be available by means of further doors or gates that can be easily opened in a manner similar to the final exit.

Marquees, tents and temporary structures

Marquees, tents and temporary structures should in general be treated in the same way as a permanent structure of the same size and layout. Some specific issues to consider include the following;

- Exit routes from marquees, tents and temporary structures may be over uneven ground or temporary flooring, duckboards, ramps, etc. so you should ensure that there are safe access and egress routes.

- Escape routes should be sited away from marquees to avoid trip hazards from guy ropes and stakes. Where necessary, barriers should be provided.

- Exits should be clearly indicated and if they consist of wall flaps they should be of a quick release design, clearly defined at the edges and so arranged as to be easily and immediately opened from the inside.

- There should be at least two exits from a marquee, and all exits should be distributed evenly around the marquee so that alternative routes are available.

More detailed recommendations are given in the *Guide to health, safety and welfare at pop concerts and similar events*.[49]

Childcare facilities/crèches

The location of childcare facilities/crèches in your premises is important since parents or guardians will often seek to return to the facility when the alarm sounds. It is therefore important that the facility is located so as to avoid parents travelling against the normal direction of escape. The childcare facility should be sited at the same level as the parents or guardians or on the route to the final exit.

4.2 Escape route layout

The examples listed in Table 3 show typical escape route solutions for a range of common building layouts. In each case the solution is for a normal risk building unless otherwise illustrated.

These are not intended to be prescriptive or exhaustive but merely to help you understand how the principles of means of escape may be applied in practice.

They are illustrative of the key features of escape route layouts and not intended to be real building layouts or to scale.

You do not need to read all of this section, you only need to consider those figures and the accompanying text which most closely resemble your premises. If your premises do not resemble these then you should seek advice from a competent person. These examples are intended to represent your existing layout; they are not to be used as design guidance.

In all of these examples the following basic principles apply:

- The farthest point on any floor to the final exit or storey exit to a protected stairway is within the overall suggested travel distance (see Table 2).

- The route to and the area near the exit is kept clear of combustibles and obstructions.

- The fire-resisting stairway is kept clear of combustibles and obstructions.

- The escape route leads to a final exit.

- Where the stairway is not a protected stairway, the final exit is visible and accessible from the discharge point of the stairway at ground floor level.

- High-risk rooms do not generally open directly into a protected stairway.

- If your fire risk assessment shows that people using any floor would be unaware of a fire you may require additional fire-protection measures, e.g. an automatic fire-detection and warning system.

- There should be more than one escape route from all parts of the premises (rooms or storeys) except for areas or storeys with an occupancy of less than 60. The figure of 60 can be varied in proportion to the risk, for a lower risk there can be a slight increase, for a higher risk, lower numbers of persons should be allowed.

Table 3: Typical examples of escape route layouts

Single-storey buildings or the ground floor of a larger building	
Ground floor with a single exit (including a mezzanine)	See Figure 39
Ground floor with a single exit (escape via a lobby)	See Figure 40
Ground floor with more than one exit	See Figure 41
Multi-storey buildings with a more than one stairway	
Two-storey, lower risk premises (e.g. church hall)	See Figure 42
Two-storey, higher risk premises (e.g. nightclub)	See Figure 43
Tall building with a firefighting shaft	See Figure 44
Multi-storey buildings with a single stairway	
Two-storey, ground and one upper floor	See Figure 45
Two-storey, basement and ground floor	See Figure 46
Three-storey, small basement, ground and one upper floor	See Figure 47
Three-storey, basement, ground and one upper floor	See Figure 48
Four-storey, ground and up to three upper floors – protected by lobbies/corridors	See Figure 49
Four-storey, ground and up to three upper floors – protected with automatic fire detection	See Figure 50

If you do not have any of the stairway configurations given, and depending on the outcome of your fire risk assessment it may be that you can achieve an equivalent level of safety by other means.

The green arrows on the Figures 39–50 represent the travel distances given in Table 2 (page 65) which should be applied.

If your building has more than ground and three upper storeys, ask advice from a competent person.

Single-storey building (or the ground floor of a larger building)

Ground floor with a single exit (including a mezzanine)

Part of your premises may have only a single exit. The example shown in Figure 39 will be generally acceptable provided that the part of the premises served only by a single exit (i.e. ground floor and mezzanine) accommodates no more than 60 people in total.

If your fire risk assessment shows that people using the mezzanine would be unaware of a fire, it may require additional fire-protection measures, e.g. an automatic fire-detection and warning system.

Note: A mezzanine covering more than half of the floor area may need to be treated as a separate floor (see two-storey buildings.

Figure 39: Ground floor, with a single exit (including a mezzanine)

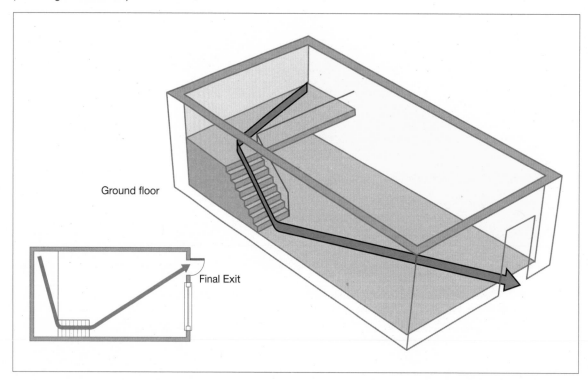

Ground floor

Final Exit

Ground floor with a single exit (escape via a lobby)

Figure 40 shows an acceptable layout for a premises with one exit, with escape via a lobby.

Restrict the total capacity to 60 persons.

Figure 40: Ground floor with a single exit (escape via a lobby)

Ground floor with more than one exit

Figure 41 shows an acceptable example of a premises with more than one exit. The figure shows acceptable examples of both cellular and open plan layouts.

Figure 41: Ground floor with more than one exit

Cellular layout

Final Exit

Final Exit

Open-plan layout

Final Exit

Final Exit

45°+

Multi-storey buildings with more than one stairway

Two-storey

If your premises has a ground floor and one upper floor and these are served by more than one stairway, it is important to understand that you may not be able to meet the suggested travel distance to a final exit (see Table 2 on page 65). In this case, stairways may therefore need to be protected by a fire-resisting enclosure as shown.

The layout shown in Figures 42 and 43 will be generally acceptable as long as the furthest point on each of your floors to the storey exit (or to a final exit using an unprotected stair) is within the overall suggested travel distance (see Table 2 on page 65).

This principle applies to taller buildings (up to 18m). However, where your building has more than three upper storeys ask advice from a competent person.

Figure 42: Two-storey, lower risk (e.g. church hall)

Figure 43: Two-storey, higher risk (e.g. nightclub)

Tall building with a firefighting shaft

Figure 44 shows a multi-storey building more than 18m high fitted with a firefighting shaft which is required for specific types of buildings. If the premises you occupy are situated in a building like this, you should ask the advice of a competent person. Further information may be found in BS 5588-5[66] and Approved Document B.[24]

Figure 44: Tall building with a firefighting shaft

Firefighting lift

SC

SC

SC

Floors higher than 18m

Upper floor

SC

SC

SC

Firefighting lift

SC

Maximum distance 18m

Firefighting stairs

Fire door

Fire-resisting wall

SC   Self-closing

Multi-storey buildings with a single stairway

Smaller places of public assembly may be provided with a single stairway. The single stairway examples provided in this section may be suitable provided that the qualifying provisions are met. If in doubt you should seek advice from a competent person.

Two-storey, ground and one upper floor

In premises with a ground floor and one upper storey with a single stairway and a simple layout, an open stairway is acceptable as indicated in Figure 45 provided that:

- The upper floor should accommodate no more than 60 people.

- The furthest point on each of the floors to the final exit is within the overall suggested travel distance for escape in one direction only (see Table 2 on page 65).

Figure 45: Two-storey, ground and one upper floor with a single stairway

First floor

Ground floor

Final Exit

Two-storey, basement and ground floor
In premises with a basement and a simple layout, an open stairway is acceptable as indicated in Figure 46 provided:

- The basement should accommodate no more than 60 people.

- The furthest point on each floor to the final exit is within the overall suggested travel distance (see Table 2 on page 65).

Figure 46: Two-storey, basement and ground floor with a single stairway

Three-storey, small basement, ground and one upper floor

In premises with a ground floor and a small basement and first floor, each served by a single stairway, the layout in Figure 47 will be generally acceptable as long as the following apply:

- The basement should accommodate no more than 60 people.

- The first floor should accommodate no more than 60 people.

- The farthest point in the basement to the door to the nearest stairway is within the overall suggested travel distance (see Table 2 on page 65).

- The stairways from the basement to ground floor level is enclosed by fire-resisting construction and leads to a final exit.

- The farthest point on the first floor to the **final** exit is within the overall suggested travel distance.

Figure 47: Three-storey, small basement, ground and one upper floor

First floor

Ground floor

Final Exit

Basement

⌐| Fire door
▬   Fire-resisting wall
SC   Self-closing

Three-storey, basement, ground and one upper floor

In premises with a basement, ground and first floor, served by a single stairway, the layout shown in Figure 48 will be generally acceptable as long as the following apply:

- The basement should accommodate no more than 60 people.

- The first floor should accommodate no more than 60 people.

- To overcome the restriction of travel distance the stairway has been completely enclosed in 30-minute fire-resisting construction and all doors onto the stairway are self-closing fire doors.

- The furthest point on each floor to the storey exit is within the overall suggested travel distance (see Table 2 on page 65).

- Where the building incorporates a basement, any stairway from the basement is separated by a fire-resisting lobby or corridor between that basement and the protected stairway.

Figure 48: Three-storey, basement, ground and one upper floor

First floor

Ground floor

Final Exit

Basement

	Fire door
	Fire-resisting wall
SC	Self-closing

Four-storey, ground and up to three upper floors

Because of the higher risk posed by premises used as a place of assembly with a single stairway to parts of the premises, and to protect the escape route by preventing smoke from entering the stairway, a protected lobby or corridor approach between the stairway and all floors (other than the top floor) should be provided as shown.

If your premises have a ground floor and up to three upper storeys and are served by a single stairway, it is important to understand that you are unlikely to be able to meet the suggested travel distance to a final exit (see Table 2 on page 65). In these circumstances it is necessary to protect the stairway by a fire-resisting enclosure as shown.

The layout shown in Figure 49 will be generally acceptable as long as the following apply:

- The upper floors should each accommodate no more than 60 people.

- The furthest point on each floor to the storey exit is within the overall suggested travel distance (see Table 2 on page 65).

- When a protected lobby or corridor approach to the stairway is employed, the travel distance is measured to the storey exit and not to the door to the lobby or corridor.

Exceptionally for very low risk premises (where densities are low, occupants are familiar with the premises, there is excellent visual awareness and very limited combustibles) automatic fire detection on all floors may be used instead of protected lobbies or corridors (see Figure 50); however, the stairway must still be protected.

Figure 49: Four-storey, ground and up to three upper floors – protected by lobbies/corridors

Top floor

SC

First and second floors

W.C.
W.C.
SC
SC
SC

Ground floor

SC
SC
SC

Final Exit

Fire door

Fire-resisting wall

SC   Self-closing

Figure 50: Four-storey, ground and up to three upper floors – protected with automatic fire detection

Top floors

First and second floors

Ground floor

Final Exit

⌐	Fire door
▬	Fire-resisting wall
SC	Self-closing
●	Automatic fire detection

Section 5 Further guidance on emergency escape lighting

The primary purpose of emergency escape lighting is to illuminate escape routes, but it also illuminates safety equipment.

The size and type of your premises and the risk to the occupants will determine the complexity of the emergency escape lighting required. Borrowed lighting may be suitable in simple premises (e.g. a small single-storey open-plan hall) where light is from a dependable source (e.g. from street lamps) and it will adequately illuminate escape routes in areas not used by the public. Where borrowed lighting is not suitable (e.g. in a high-risk area such as a kitchen), then a number of torches in strategic positions for the use of staff only, could be considered.

In simple premises where borrowed lighting is not suitable in areas used by the public, single 'stand-alone' escape lighting units may be sufficient and these can sometimes be combined with exit or directional signs. The level of general illumination should not be significantly reduced by the sign.

In more complex premises, e.g. those that are part of a multi-storey or multi-occupied building, a more comprehensive emergency escape lighting system may be appropriate.

You will have identified the escape routes when carrying out your fire risk assessment and need to ensure that they are all adequately lit. If there are escape routes that are not permanently illuminated by normal lighting, such as external stairs, then a switch, clearly marked 'Escape lighting', or some other means of switching on the lighting should be provided at the entry to that area/stairs.

An emergency escape lighting system should normally covers the following:

- each exit door;
- escape routes;
- intersections of corridors;
- outside each final exit and on external escape routes;
- emergency escape signs;
- stairways so that each flight receives adequate light;
- changes in floor level;
- windowless rooms and toilet accommodation exceeding 8m²;
- firefighting equipment;
- fire alarm call points;
- equipment that would need to be shut down in an emergency;
- lifts; and
- halls or other areas greater than 60m².

It is not necessary to provide individual lights (luminaires) for each item above, but there should be a sufficient overall level of light to allow them to be visible and usable.

Emergency escape lighting can be both 'maintained', i.e. on all the time, or 'non-maintained' which only operates when the normal lighting fails. Systems or individual lighting units (luminaires) are designed to operate for durations of between one and three hours. In practice, the three-hour units are the most popular and can help with maintaining limited continued use of your premises during a power failure (other than in an emergency situation).

Figure 51: Luminaires

Emergency escape lighting (luminaires) can be stand-alone dedicated units or incorporated into normal light fittings. There are highly decorative versions of these for those areas that demand aesthetically pleasing fixtures. Power supplies can be rechargeable batteries integral to each unit, a central battery bank or an automatic start generator.

To complement emergency escape lighting, people, especially those unfamiliar with the premises, can be helped to identify exit routes by the use of way-guidance equipment. Way-guidance systems usually comprise photo-luminescent material, lines of LEDs, or strips of miniature incandescent lamps, forming a continuous marked escape route at lower level (Figure 52). These systems have proved particularly effective when people have had to escape through smoke, including for partially-sighted people. They can be particularly useful in premises where they can provide marked routes on floors and in multi-storey premises they can direct people to escape routes which are seldom used.

If you decide that you need to install emergency escape lighting or to modify your existing system, any work should be carried out by a competent person in accordance with the appropriate standards.

Figure 52: A 'way-guidance' system

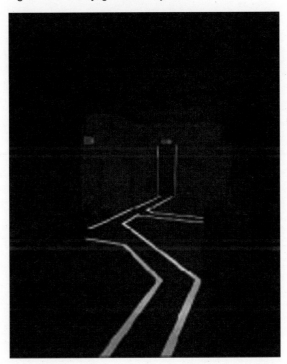

Further guidance is given in BS 5266-1[27A] and BS 5266-8.[27]

Emergency escape lighting by gas or other means not using electricity

In premises where escape lighting by gas or other means not using electricity is already installed, the escape lighting should be equivalent to that powered by electricity. Appropriate safety measures should be provided, for example with regards to fixing, location of combustible materials, protection from draughts and emergency shut-off.

Lighting for special effects

Lasers, strobes or other form of high intensity lighting are not suitable for emergency use.

Lighting conditions for stair treads

Where lighting conditions render it desirable the nosings of all stair treads should be made conspicuous by painting or other effective means. Guidance is available in BS 8300.[14]

Maintenance and testing of emergency escape lighting

All emergency escape lighting systems should be regularly tested and properly maintained to an appropriate standard. Most existing systems will need to be manually tested. However, some modern systems have self-testing facilities that reduce routine checks to a minimum.

Depending on your type of installation you should be able to carry out most of the routine tests yourself. The test method will vary. If you are not sure how to carry out these tests you should contact your supplier or other competent person.

Figure 53: A test key

Test facilities often take the form of a 'fishtail' key (see Figure 53) inserted in a special switch either near the main fuse board or adjacent to relevant light switches.

Typically, testing would include:

- a daily visual check of any central controls;

- a monthly function test by operating the test facility for a period sufficient to ensure that each emergency lamp illuminates; and

- an annual full discharge test.

Particular care needs to be taken following a full discharge test. Batteries typically take 24 hours to re-charge and the premises should not be re-occupied until the emergency lighting system is fully functioning unless alternative arrangements have been made.

It is good practice to keep a record of tests.

See BS 5266-8.[27] for more information.

Section 6 Further guidance on signs and notices

Escape signs

In simple premises, a few signs indicating the alternative exit(s) might be all that is needed. In larger multi-room premises, a series of signs directing people along the escape routes towards the final exit might be needed.

Many people with poor vision retain some sight and are able to recognise changing or contrasting colour to provide them with visual clues when moving around a building.* It may be sufficient to paint any columns and walls in a contrasting colour and to highlight changes in level by, for example, making the nosing to step and stair treads a contrasting colour (see BS 8300[14] for further guidance).

For people with no sight, a well-managed 'buddy system', continuous handrails, a sound localisation system (which helps people to move towards an alert sound or the installation of more tactile aids may be appropriate.

People will usually attempt to leave premises by the same way that they entered. As these entrances in places of assembly will then be the exits of choice it follows that the other exit facilities provided for the premises may not be fully utilised. For this reason all available exits should be clearly indicated so that the public are aware that there are ways to leave the building other than by the doors which they used to gain admission. In addition, the provision of well sign-posted exits in full view of an audience can give a feeling of security in an emergency situation.

Maintained internally illuminated exit signs will be required where the lighting may be dimmed or extinguished (e.g. lecture rooms).

Exit signs should be clearly visible whenever the public are present.

Positioning of escape route signs

The presence of other signs in premises (such as advertising) can distract attention from, or obscure the visibility of, escape signs. This could affect people's ability to see and understand escape signs, particularly if there is a fire evacuation. Always ensure that escape signs are not overwhelmed.

---

* The Royal National Institute of the Blind estimates that only about 4% of visually impaired people are totally blind.

Escape signs should meet the following criteria:

- They should provide clear, unambiguous information to enable people to safely leave a building in an emergency.

- Every escape route sign should, where necessary, incorporate, or be accompanied by, a directional arrow. Arrows should not be used on their own.

- If the escape route to the nearest exit is not obvious then it should be indicated by a sign(s).

- Signs should be positioned so that a person escaping will always have the next escape route sign in sight.

- Escape signs should be fixed above the door in the direction of escape and not be fixed to doors, as they will not be visible if the door is open.

- Signs mounted above doors should be at a height of between 2.0m and 2.5m above the floor.

- Signs on walls should be mounted between 1.7m and 2.0m above the floor.

- Mounting heights greater than 2.5m may be used for hanging signs, e.g. in large open spaces or for operational reasons, but care should be taken to ensure that such signs are both conspicuous and legible. In such case larger signs may be necessary.

- Signs should be sited at the same height throughout the escape route, so far as is reasonably practicable.

Escape sign design

For a sign to comply with signs and signals regulations it must be pictographic (see Figures 54 and 55). The pictogram can be supplemented by text if this is considered necessary to make the sign easily understood (BS-type sign), but you must not have a fire safety sign that uses only text. Either type of sign can be used but different types should not be mixed. Appropriate signs should take into account the needs of those who may need to use them.

The legibility of escape signs is determined by the size of the sign, its level of illumination and the distance over which it is viewed. The use of signs within the same premises should follow a consistent design pattern or scheme. You should not rely on a few outsized signs which may encourage people to travel to a particular escape route when other more appropriate routes should be used.

In multi-occupied premises, co-operation between the respective responsible persons should be sought to ensure that, as far as possible, all signs in the building conform to a single pattern or scheme.

Figure 54: BS-type sign

Figure 55: Euro sign

Other safety signs and notices
A number of other mandatory signs such as 'Fire action' notices may also be necessary.

Fire doors that have been fitted with self-closing devices should be labelled 'Fire door – keep shut' on both sides (see Figure 56). Fire-resisting doors to cupboards, stores and service ducts that are not self-closing because they are routinely kept locked should be labelled 'Fire door – keep locked' on the outside.

Figure 56: Fire door 'keep shut' notice

Signs should indicate non-automatic fire safety equipment if there is any doubt about its location, e.g. fire extinguishers that are kept in cabinets or in recesses.

A notice with the words 'Push bar to open' should be permanently displayed immediately above the push-bar on all doors fitted with a panic bolt or panic latch.

A notice with the words 'Fire escape – keep clear' should be permanently displayed at about eye level on the external face of all doors which are provided as a means of escape in case of fire and which, because they are not normally used, may become obstructed.

Staff notices
In simple premises where there are a limited number of escape routes, it may be reasonable to provide staff with verbal reminders of what they need to do if there is a fire. In some premises, you could consider providing a short written statement that could, for example, be delivered with staff pay slips every six months.

Where there is a high turnover of staff, a more considered approach for staff notices and instructions will be necessary. As well as positioning the fire instructions notice on escape routes adjacent to fire break-glass call points (see Figure 57), put them where staff frequently assemble in the premises, e.g. the canteen and locker rooms.

In some premises, such as nightclubs, consider whether it is suitable for fire instruction notices to be available in public areas or whether they should only be in areas to which staff have access.

Figure 57: Fire action notice

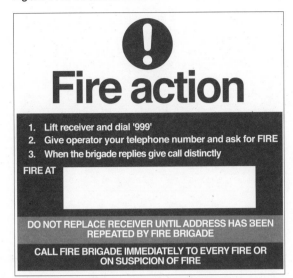

If your premises are routinely expected to accommodate people whose first language is not English you may need to consider providing instruction in more than one language. The interpretation should always convey an identical message.

Illumination

All signs and notices will need illumination to ensure they are conspicuous and legible. There are a number of options available to achieve this, such as:

- external illumination; and

- internal illumination.

The supplier or other competent person can give you further advice.

All exit and directional signs indicating the exits from any part of the premises to which the public are admitted should (unless they are self-luminous fire safety signs) be illuminated by means of the normal lighting and the emergency escape lighting at all times when the public are on the premises.

Signs or notices of the photo-luminescent type, i.e. where the active material making up the luminous parts of such signs or notices needs a period of exposure to light before they become visible in darkness (but get fainter with time), are not a substitute for appropriate emergency lighting and should only be used where other forms of illumination are present.

Further guidance

Further guidance on fire safety signs can be found in BS 5499-5[29] and BS 5499-4.[29A] Published guidance[5,6] on compliance with health and safety legislation on signs is also available. Guidance about the use of photo-luminescent fire safety signs and notices can be found in BS 5266-6.[78]

Section 7 Further guidance on recording, planning, informing, instructing and training

7.1 Fire safety records

Keeping up-to-date records of your fire risk assessment can help you effectively manage the fire strategy for your premises and demonstrate how you are complying with fire safety law.

Even if you do not have to record the fire risk assessment, it can be helpful to keep a record of any co-operation and exchange of information made between employers and other responsible people for future reference.

It is best to keep a dedicated record of all maintenance of fire-protection equipment and training. There is no one 'correct' format specified for this. Suitable record books are available from trade associations and may also be available from your local enforcing authority.

In all cases the quality of these records may also be regarded as a good indicator of the overall quality of the safety management structure.

Your records should be kept in a specified place on the premises (for example, in the management's office), and should include:

- details of any significant findings from the fire risk assessment and any action taken (see Part 1, Section 4.1);

- testing and checking of escape routes, including final exit locking mechanisms, such as panic devices, emergency exit devices and any electromagnetic devices;

- testing of fire warning systems, including weekly alarm tests and periodic maintenance by a competent person;

- recording of false alarms;

- testing and maintenance of emergency lighting systems;

- testing and maintenance of fire extinguishers, hose reels and fire blankets, etc.;

- if appropriate, testing and maintenance of other fire safety equipment;

- recording and training of relevant people and fire evacuation drills;

- planning, organising, policy and implementation, monitoring, audit and review;

- maintenance and audit of any systems that are provided to help the fire and rescue service;

- the arrangements in a large multi-occupied building for a co-ordinated emergency plan or overall control of the actions you or your staff should take if there is a fire; and

- all alterations, tests, repairs and maintenance of the fire safety systems, including passive systems such as fire doors.

Other issues that you may wish to record include:

- the competence, qualifications and status of the persons responsible for carrying out inspections and tests;

- the results of periodic safety audits, reviews, inspections and tests, and any remedial action taken;

- all incidents and circumstances which had the potential to cause accidents and monitor subsequent remedial action; and

- a record of the building use, the fire prevention and protection measures in place and high-risk areas.

You should ensure that no other management decisions or policies compromise safety.

Your documentation should be available for inspection by representatives of the enforcing authority.

More detailed advice is given in BS 5588-12.[51]

Figure 58 is an example of how to record some individual stages of the process in more detail. A blank version of this form is provided in Appendix A2.

Figure 58: Example record of significant findings

Risk Assessment – Record of significant findings

Risk assessment for | Assessment undertaken by

Company	Inspirations		Date	02/02/2006
Address	The Avenue		Completed by	J Brown
	Old Town		Signature	J Brown
	AB12 3CD			

Sheet number	Floor/area	Use
One	First floor	Art Gallery

Step 1 – Identify fire hazards

Sources of ignition	Sources of fuel	Sources of oxygen
Display lighting	Exhibitions	No additional sources
Electrical heaters	Display material	
Staff tea making equipment	Records etc. in office	

Step 2 – People at risk

3 Permanent staff	50 members of the public
2 temporary staff	(peak times)

Step 3 – Evaluate, remove, reduce and protect from risk

(3.1) Evaluate the risk of the fire occuring	• Display lighting in close proximity to exhibits • Floor mounted heaters – combustible materials in close proximity • Kettle and toaster in tea area in rear office adjacent to substantial quantities of records/unfiled papers
(3.2) Evaluate the risk to people from a fire starting in the premises	• Gallery area predominantly open plan • Fire in rear office/tea area could go undetected and present a hazard to the gallery • Fire in ground floor gift shop could obstruct stairway to the gallery
(3.3) Remove and reduce the hazards that may cause a fire	• Reposition display lighting • Wall mount electrical heaters • Remove toaster – replace kettle • Records etc. to be filed in metal filing cabinets in office
(3.4) Remove and reduce the risks to people from a fire	The current fire precautions (refer to floor plan) have been assessed with respect to the findings of the risk assessment and are considered adequate with the following exceptions: • Replace office door with a fire door to seperate office from gallery • Provide fire exit sign to the external escape staircase from office • Replace self closer to the door protecting front staircase from shop at ground floor level

Assessment review

Assessment review date	Completed by	Signature

Review outcome (where substantial changes have occurred a new record sheet should be used)

Notes:
(1) The risk assessment record significant findings should refer to other plans, records or other documents as necessary.
(2) The information in this record should assist you to develop an emergency plan; coordinate measures with other 'responsible persons' in the building; and to inform and train staff and inform other relevant persons.

In premises with 'engineered fire safety strategies', a fire policy manual should be provided in addition to any other records. Guidance on the structure of fire engineering policy manuals is given in BS 7974-0 Section 5: Reporting and presentation.[30]

Fire safety audit

A fire safety audit can be used alongside your fire risk assessment to identify what fire safety provisions exist in your premises.

When carrying out a review of your fire safety risk assessment, a pre-planned audit can quickly identify if there have been any significant changes which may affect the fire safety systems and highlight whether a full fire risk assessment is necessary.

Plans and specifications

Plans and specifications can be used to assist understanding of a fire risk assessment or emergency plan. Even where not needed for this purpose they can help you and your staff keep your fire risk assessment and emergency plan under review and help the fire and rescue service in the event of fire. Any symbols used should be shown on a key. Plans and specifications could include the following:

- essential structural features such as the layout of function rooms, escape doors, wall partitions, corridors, stairways, etc. (including any fire-resisting structure and self-closing fire doors provided to protect the escape routes);

- location of refuges and lifts that have been designated suitable for use by disabled people and others who may need assistance to escape in case of a fire;

- methods for fighting fire (details of the number, type and location of the firefighting equipment);

- location of manually-operated fire alarm call points and control equipment for fire alarms;

- location of any emergency lighting equipment and the exit route signs;

- location of any high-risk areas, equipment or process that must be immediately shut down by staff on hearing the fire alarm;

- location of any automatic firefighting systems, risers and sprinkler control valves;

- location of the main electrical supply switch, the main water shut-off valve and, where appropriate, the main gas or oil shut-off valves; and

- plans and specifications relating to all recent constructions.

This information should be passed on to any later users or owners of the premises.

7.2 Emergency plans

Emergency plan and contingency plans

Your emergency plan should be appropriate to your premises and could include:

- how people will be warned if there is a fire;

- what staff should do if they discover a fire;

- how the evacuation of the premises should be carried out;

- identification of key escape routes;

- arrangements for fighting fire;

- arrangements for the safe evacuation of people identified as being especially at risk, such as those with disabilities and children;

- contingency plans, e.g. restrictions on the use of the building, for when life safety systems are out of order, e.g. evacuation lifts, fire-detection and warning systems;

- how the fire and rescue service and any other necessary services will be called and who will be responsible for doing this; and

- what training employees need and the arrangements for ensuring that this training is given.

As part of your emergency plan it is good practice to prepare post-incident plans for dealing with situations that might arise such as those involving:

- unaccompanied children;

- people with personal belongings (especially valuables) still in the building;

- people in a state of undress (e.g. indoor sportswear);

- people wishing to rejoin friends;

- getting people away from the building (e.g. to transport); and

- inclement weather.

You should also assess the risk of any incident occurring which might prejudice public safety or disrupt normal operations, for example, power cuts, bomb threats or crowd disorder. Such incidents often arise with little or no warning and may not be capable of being dealt with by the management operating under normal circumstances.

You should therefore prepare contingency plans to determine specific actions and/or the mobilisation of specialist resources.

Guidance on developing health and safety managment policy has been published by the HSE.[31]

Responsibilities for short-term hiring or leasing and for shared use

It is crucial that the temporary responsible person understands their duties for the duration of the event or function. Figure 59 gives an example of the sorts of recommendations that could be provided by the owner of the premises to the temporary responsible person, who must be made aware of their legal duties.

Figure 59: Emergency plan for the temporary responsible person

As the responsible person for the event/ function, etc. you have legal duties with regards to the safety of those persons assisting or attending the event.

Before the event or function you should be aware of:

- what fire protection systems are present;

- how a fire will be detected;

- how people will be warned if there is a fire;

- what staff should do if they discover a fire;

- how the evacuation of the premises should be carried out;

- where people should assemble after they have left the premises and procedures for checking whether the premises have been evacuated;

- identification of key escape routes and exits, how people can gain access to them and escape to a place of safety;

- arrangements for fighting fire;

- specific arrangements, if necessary, for high-fire-risk areas;

- how the fire and rescue service and any other necessary services will be called;

- procedures for meeting the fire and rescue service on their arrival and notifying them of any special risks, e.g. the location of highly flammable materials;

- what instruction employees or helpers need and the arrangements for ensuring that this training is given;

- limitation on numbers of people;

- any chains and padlocks which need removing from fire exits, etc.;

- exit doors which are required to be in the open position are secure; and

- checking that all escape routes are clear of obstructions and combustibles.

Before the event or function you should decide:

- the arrangements for fighting fire;

- the arrangements for means of escape for disabled persons;

- the duties and identity of staff who have specific responsibilities if there is a fire;

- the arrangements for the safe evacuation of people identified as being especially at risk, such as contractors, those with disabilities, children, members of the public and visitors;

- how you will proceed if life safety systems are out of order, e.g. fire-detection and alarm systems, sprinklers or smoke control systems;

- who will be responsible for calling the fire and rescue service and any other necessary services;

- who will meet the fire and rescue service on their arrival and notifying them of any special risks, e.g. the location of highly flammable materials; and

- your plans to deal with people once they have left the premises, especially children.

At the start of the event or function you should notify all those present about:

- the smoking policy;

- the emergency warning signal;

- who is supervising and how to identify them;

- location of exits and escape routes;

- taking only valuables immediately to hand but not to go to collect other belongings;

- the location of muster points; and

- what will happen after that (e.g. re-entry to the building).

During the event or function you should ensure that:

- escape routes and exits do not become blocked;

- your smoking policy is adhered to;

- no naked flames are started (unless authorised e.g. candles);

- where naked flames are present that combustible material is kept clear;

- rooms do not become overcrowded;

- noise levels cannot drown out the fire alarm; and

- if necessary, the number of persons in your premises is limited or controlled.

The above list is for example only and will need to be specific to any particular venue or premises.

7.3 Information, instruction, co-operation and co-ordination

Supplying information

You must provide easily understandable information to employees, the parents of children you may employ, and to employers of other persons working in your premises about the measures in place to ensure a safe escape from the building and how they will operate, for example:

- any significant risks to staff and other relevant persons that have been identified in your fire risk assessment or any similar assessment carried out by another user and responsible person in the building;

- the fire prevention and protection measures and procedures in your premises and where they impact on staff and other relevant persons in the building;

- the procedures for fighting a fire in the premises; and

- the identity of people who have been nominated with specific responsibilities in the building.

Even if you do not have to record the fire risk assessment, it would be helpful to keep a record of any co-operation and exchange of information made between employers and other responsible people for future reference.

You need to ensure that all staff and, where necessary, other relevant persons in the building, receive appropriate information in a way that can be easily understood. This might include any special instructions to particular people who have been allocated a specific task, such as shutting down equipment or guiding people to the nearest exit.

Duties of employees to give information

Employees also have a duty to take reasonable care for their own safety and that of other people who may be affected by their activities. This includes the need for them to inform their employer of any activity that they consider would present a serious and immediate danger to their own safety and that of others.

Dangerous substances

HSE publishes guidance[8] about specific substances where appropriate information may need to be provided. If any of these, or any other substance that is not included but nevertheless presents more than a slight risk, is present in your premises, then you must provide such information to staff and others, specifically you must:

- name the substance and the risks associated with it, e.g. how to safely use or store the product to avoid creating highly flammable vapours or explosive atmospheres;

- identify any legislative provisions that may be associated with the substance;

- allow employees access to the hazardous substances safety data sheet; and

- inform the local fire and rescue service where dangerous substances are present on the premises.

Information to the emergency services

In addition to providing information to the fire and rescue service when dangerous substances are present in sufficient quantities to pose an enhanced risk, it will also be helpful to inform them of any short term changes that might have an impact on their firefighting activities; e.g. in the event of temporary loss of a firefighting facility and temporary alterations.

Procedures should also include meeting and briefing the fire and rescue service when they arrive.

Instruction

You will need to carefully consider the type of instructions to staff and other people working in your premises. Written instructions must be concise, comprehensible and relevant and therefore must be reviewed and updated as new working practices and hazardous substances are introduced.

Inclusive access and employment policies mean that people with learning difficulties may now be present in a range of premises and your fire risk assessment should consider whether further instruction or guidance is necessary to ensure that your evacuation strategy is appropriate and understood by everyone.

Instructions will need to be given to people delegated to carry out particular tasks, for example:

- removing additional security, bolts, bars or chains on final exit doors before the start of business to ensure that escape routes are accessible;

- daily, weekly, quarterly and yearly checks on the range of fire safety measures (in some premises some of the work may be contracted out to a specialist company);

- safety considerations when closing down the premises at the end of the day, e.g. removing rubbish, ensuring enough exits are available for people that remain and closing fire doors and shutters;

- leaving hazardous substances in a safe condition when evacuating the building; and

- the safe storage of hazardous substances at the end of the working day.

Specific instructions may be needed about:

- how staff will help members of the public/visitors to leave the building;

- 'sweeping' of the premises floors by staff to guide people to the nearest exit when the fire alarm sounds;

- designating particular areas of your premises to check that no one remains inside;

- calling the emergency services;

- carrying out evacuation roll calls;

- taking charge at the assembly area;

- meeting and directing fire engines; and

- cover arrangements when nominated people are on leave.

Co-operation and co-ordination

Where you share premises with others (this includes people who are self-employed or in partnership), each responsible person, i.e. each employer, owner or other person who has control over any part of the premises, will need to co-operate and co-ordinate the findings of their separate fire risk assessments to ensure the fire precautions and protection measures are effective throughout the building. This could include:

- co-ordinating an emergency plan (see Section 7.2 for features of an emergency plan);

- identifying the nature of any risks and how they may affect others in or about the premises;

- identifying any fire-prevention and protection measures;

- identifying any measures to mitigate the effects of a fire; and

- arranging any contacts with external emergency services, and calling the fire and rescue service.

7.4 Fire safety training

Staff training
The actions of staff if there is a fire are likely to be crucial to their safety and that of other people in the premises. All staff should receive basic fire safety induction training and attend refresher sessions at pre-determined intervals.

You should ensure that all staff and contractors are told about the emergency plan and are shown the escape routes.

The training should take account of the findings of the fire risk assessment and be easily understood by all those attending. It should include the role that those members of staff will be expected to carry out if a fire occurs.

In addition to the guidance given in Part 1, Step 4.4, as a minimum all staff should receive training about:

- the items listed in your emergency plan;

- the importance of fire doors and other basic fire-prevention measures;

- where relevant, the appropriate use of firefighting equipment;

- the importance of reporting to the assembly area;

- exit routes and the operation of exit devices, including physically walking these routes;

- general matters such as permitted smoking areas or restrictions on cooking other than in designated areas; and

- assisting disabled persons where necessary.

Training is necessary:

- when staff start employment or are transferred into the premises;

- when changes have been made to the emergency plan and the preventive and protective measures;

- where working practices and processes or people's responsibilities change;

- to take account of any changed risks to the safety of staff or other relevant persons;

- to ensure that staff know what they have to do to safeguard themselves and others on the premises;

- where staff are expected to assist disabled persons; and

- if member of staff may take on the role of duty manager.

Training should be repeated as often as necessary and should take place during working hours.

Whatever training you decide is necessary to support your fire safety strategy and emergency plan, it should be verifiable.

Enforcing authorities may want to examine records as evidence that adequate training has been given.

Fire marshals

Staff expected to undertake the role of fire marshals (often called fire wardens) would require more comprehensive training. Their role may include:

- helping those on the premises to leave;

- checking the premises to ensure everyone has left;

- using firefighting equipment if safe to do so;

- liaising with the fire and rescue service on arrival;

- shutting down vital or dangerous equipment; and

- performing a supervisory/managing role in any fire situation.

Training for this role may include:

- detailed knowledge of the fire safety strategy of the premises;

- awareness of human behaviour in fires;

- how to encourage others to use the most appropriate escape route;

- how to search safely and recognise areas that are unsafe to enter;

- the difficulties that some people, particularly if disabled, may have in escaping and any special evacuation arrangements that have been pre-planned;

- additional training in the use of firefighting equipment;

- an understanding of the purpose of any fixed firefighting equipment such as sprinklers; and

- reporting of faults, incidents and near misses.

For licensed premises, door supervisors may have a fire safety function and appropriate training should be provided.

Fire drills

Once the emergency plan has been developed and training given, you will need to evaluate its effectiveness. The best way to do this is to perform a fire drill. This should be carried out at least annually or as determined by your fire risk assessment. If you have a high staff turnover, you may need to carry them out more often.

A well-planned and executed fire drill will confirm understanding of the training and provide helpful information for future training. The responsible person should determine the possible objectives of the drill such as to:

- identify any weaknesses in the evacuation strategy;

- test the procedure following any recent alteration or changes to working practices;

- familiarise new members of staff with procedures; and

- test the arrangements for disabled people.

Who should take part?

Within each building the evacuation should be for all occupants except those who may need to ensure the security of the premises, or people who, on a risk-assessed basis, are required to remain with particular equipment or processes that cannot be closed down.

Where appropriate, you may find it helpful to include members of the public in your fire drill – ensuring that all necessary health and safety issues are addressed before you do so.

Carrying out the drill

For premises that have more than one escape route, the escape plan should be designed to evacuate all people on the assumption that one exit or stairway is unavailable because of the fire. This could be simulated by a designated person being located at a suitable point on an exit route. Applying this scenario to different escape routes at each fire drill will encourage individuals to use alternative escape routes which they may not normally use.

When carrying out the drill you might find it helpful to:

- circulate details concerning the drill and inform all staff of their duty to participate. It may not be beneficial to have 'surprise drills' as the health and safety risks introduced may outweigh the benefits;

- ensure that equipment can be safely left;

- nominate observers;

- inform the alarm receiving centre if the fire-warning system is monitored (if the fire and rescue service is normally called directly from your premises, ensure that this does not happen);

- inform visitors and members of the public if they are present; and

- ask a member of staff at random to set off the alarm by operating the nearest alarm call point using the test key. This will indicate the level of knowledge regarding the location of the nearest call point.

More detailed information on fire drills and test evacuations are given in BS 5588-12.[51]

The roll call/checking the premises have been evacuated

In some cases it may be appropriate to carry out a roll call as soon as possible at the designated assembly point(s), and/or receive reports from wardens designated to 'sweep' the premises. You should note any people who are unaccounted for. In a real evacuation this information will need to be passed to the fire and rescue service on arrival.

Check that people assembled at the evacuation point.

Once the roll call is complete or all reports have been received, allow people to return to the building. If the fire warning system is monitored, inform the alarm receiving centre that the drill has now been completed and record the outcomes of the drill.

Monitoring and debrief

Throughout the drill the responsible person and nominated observers should pay particular attention to:

- communication difficulties with regard to the roll call and establishing that everyone is accounted for;

- the use of the nearest available escape routes as opposed to common circulation routes;

- difficulties with the opening of final exit doors;

- difficulties experienced by people with disabilities;

- the roles of specified people, e.g. fire wardens;

- inappropriate actions, e.g. stopping to collect personal items, attempting to use lifts, etc.; and

- windows and doors not being closed as people leave.

On-the-spot debriefs are useful to discuss the fire drill, encouraging feedback from everybody. Later, reports from fire wardens and observations from people should be collated and reviewed. Any conclusions and remedial actions should be recorded and implemented.

Section 8 Quality assurance of fire protection equipment and installation

Fire protection products and related services should be fit for their purpose and properly installed and maintained in accordance with the manufacturer's instructions or a relevant standard.

Third party certification schemes for fire protection products and related services are an effective means of providing the fullest possible assurances, offering a level of quality, reliability and safety that non-certificated products may lack. This does not mean goods and services that are not third party approved are less reliable, but there is no obvious way in which this can be demonstrated.

Third party quality assurance can offer comfort, both as a means of satisfying you that goods and services you have purchased are fit for purpose, and as a means of demonstrating that you have complied with the law.

However, to ensure the level of assurance offered by third party schemes, you should always check whether the company you employ sub-contracts work to others. If they do, you will want to check that the sub-contractors are subject to the same level of checks of quality and competence as the company you are employing.

Your local fire and rescue service, fire trade associations or your own trade association may be able to provide further details about third party quality assurance schemes and the various organisations that administer them.

Appendix A

A1 Example fire safety maintenance checklist

A fire safety maintenance checklist can be used As a means of supporting your fire safety policy. This list is not intended to be comprehensive and should not be used as a substitute for carrying out a fire risk assessment.

You can modify the example, where necessary, to fit your premises and may need to incorporate the recommendations of manufacturers and installers of the fire safety equipment/systems that you may have installed in your premises.

Any ticks in the grey boxes should result in further investigation and appropriate action as necessary. In larger and more complex premises you may need to seek the assistance of a competent person to carry out some of the checks.

	Yes	No	N/A	Comments
Daily checks (not normally recorded)				
Escape routes				
Can all fire exits be opened immediately and easily?	☐	☐	☐	
Are fire doors clear of obstructions?	☐	☐	☐	
Are escape routes clear?	☐	☐	☐	
Fire warning systems				
Is the indicator panel showing 'normal'?	☐	☐	☐	
Are whistles, gongs or air horns in place?	☐	☐	☐	
Escape lighting				
Are luminaires and exit signs in good condition and undamaged?	☐	☐	☐	
Is emergency lighting and sign lighting working correctly?	☐	☐	☐	
Firefighting equipment				
Are all fire extinguishers in place?	☐	☐	☐	
Are fire extinguishers clearly visible?	☐	☐	☐	
Are vehicles blocking fire hydrants or access to them?	☐	☐	☐	
Weekly checks				
Escape routes				
Do all emergency fastening devices to fire exits (push bars and pads, etc.) work correctly?	☐	☐	☐	
Are external routes clear and safe?	☐	☐	☐	
Fire warning systems				
Does testing a manual call point send a signal to the indicator panel? (Disconnect the link to the receiving centre or tell them you are doing a test.)	☐	☐	☐	
Did the alarm system work correctly when tested?	☐	☐	☐	
Did staff and other people hear the fire alarm?	☐	☐	☐	
Did any linked fire protection systems operate correctly? (e.g. magnetic door holder released, smoke curtains drop)	☐	☐	☐	

113

	Yes	No	N/A	Comments
Weekly checks *continued*				
Do all visual alarms and/or vibrating alarms and pagers (as applicable) work?	☐	☐	☐	
Do voice alarm systems work correctly? Was the message understood?	☐	☐	☐	
Escape lighting				
Are charging indicators (if fitted) visible?	☐	☐	☐	
Firefighting equipment				
Is all equipment in good condition?	☐	☐	☐	
Additional items from manufacturer's recommendations.	☐	☐	☐	
Monthly checks				
Escape routes				
Do all electronic release mechanisms on escape doors work correctly? Do they 'fail safe' in the open position?	☐	☐	☐	
Do all automatic opening doors on escape routes 'fail safe' in the open position?	☐	☐	☐	
Are fire door seals and self-closing devices in good condition?	☐	☐	☐	
Do all roller shutters provided for fire compartmentation work correctly?	☐	☐	☐	
Are external escape stairs safe?	☐	☐	☐	
Do all internal self-closing fire doors work correctly?	☐	☐	☐	
Escape lighting				
Do all luminaires and exit signs function correctly when tested?	☐	☐	☐	
Have all emergency generators been tested? (Normally run for one hour.)	☐	☐	☐	
Firefighting equipment				
Is the pressure in 'stored pressure' fire extinguishers correct?	☐	☐	☐	
Additional items from manufacturer's recommendations.	☐	☐	☐	
Three-monthly checks				
General				
Are any emergency water tanks/ponds at their normal capacity?	☐	☐	☐	
Are vehicles blocking fire hydrants or access to them?	☐	☐	☐	
Additional items from manufacturer's recommendations.	☐	☐	☐	
Six-monthly checks				
General				
Has any firefighting or emergency evacuation lift been tested by a competent person?	☐	☐	☐	
Has any sprinkler system been tested by a competent person?	☐	☐	☐	
Have the release and closing mechanisms of any fire-resisting compartment doors and shutters been tested by a competent person?	☐	☐	☐	
Fire warning system				
Has the system been checked by a competent person?	☐	☐	☐	

	Yes	No	N/A	Comments
Six-monthly checks *continued*				
Escape lighting				
Do all luminaires operate on test for one third of their rated value?	☐	☐	☐	
Additional items from manufacturer's recommendations.	☐	☐	☐	
Annual checks				
Escape routes				
Do all self-closing fire doors fit correctly?	☐	☐	☐	
Is escape route compartmentation in good repair?	☐	☐	☐	
Escape lighting				
Do all luminaires operate on test for their full rated duration?	☐	☐	☐	
Has the system been checked by a competent person?	☐	☐	☐	
Firefighting equipment				
Has all firefighting equipment been checked by a competent person?	☐	☐	☐	
Miscellaneous				
Has any dry/wet rising fire main been tested by a competent person?	☐	☐	☐	
Has the smoke and heat ventilation system been tested by a competent person?	☐	☐	☐	
Has external access for the fire service been checked for ongoing availability?	☐	☐	☐	
Have any firefighters' switches been tested?	☐	☐	☐	
Has the fire hydrant bypass flow valve control been tested by a competent person?	☐	☐	☐	
Are any necessary fire engine direction signs in place?	☐	☐	☐	

A2 Example form for recording significant findings

Risk Assessment – Record of significant findings		
Risk assessment for	**Assessment undertaken by**	
Company	Date	
Address	Completed by	
	Signature	
Sheet number	**Floor/area**	**Use**

Step 1 – Identify fire hazards		
Sources of ignition	**Sources of fuel**	**Sources of oxygen**

Step 2 – People at risk

Step 3 – Evaluate, remove, reduce and protect from risk

(3.1) Evaluate the risk of the fire occuring

(3.2) Evaluate the risk to people from a fire starting in the premises

(3.3) Remove and reduce the hazards that may cause a fire

(3.4) Remove and reduce the risks to people from a fire

Assessment review

Assessment review date	**Completed by**	**Signature**

Review outcome (where substantial changes have occurred a new record sheet should be used)

Notes:

(1) The risk assessment record significant findings should refer to other plans, records or other documents as necessary.

(2) The information in this record should assist you to develop an emergency plan; coordinate measures with other 'responsible persons' in the building; and to inform and train staff and inform other relevant persons.

Appendix B

Technical information on fire-resisting separation, fire doors and door fastenings

B1 Fire-resisting separation

General

The materials from which your premises are constructed may determine the speed with which a fire may spread, affecting the escape routes that people will use. A fire starting in a building constructed mainly from readily combustible material will spread faster than one where modern fire-resisting construction materials have been used. Where non-combustible materials are used and the internal partitions are made from fire-resisting materials, the fire will be contained for a longer period, allowing more time for the occupants to escape.

Because of the requirements of the Building Regulations you will probably already have some walls and floors that are fire-resisting and limitations on the surface finishes to certain walls and ceilings.

You will need to consider whether the standard of fire resistance and surface finishing in the escape routes is satisfactory, has been affected by wear and tear or alterations and whether any improvements are necessary.

The following paragraphs give basic information on how fire-resisting construction can provide up to 30 minutes protection to escape routes. This is the standard recommended for most situations. If you are still unsure of the level of fire resistance which is necessary after reading this information, you should consult a fire safety expert.

Fire-resisting construction

The fire resistance of a wall or floor is dependent on the quality of construction and materials used. Common examples of types of construction that provide 30-minute fire resistance to escape routes if constructed to the above standards are:

- internal framed construction wall, non-load bearing, consisting of 72mm x 37mm timber studs at 600mm centres and faced with 12.5mm of plasterboard with all joints taped and filled (see Figure 60);

- internal framed construction, non-load bearing, consisting of channel section steel studs at 600mm centres faced with 12.5mm of plasterboard with all joints taped and filled; and

- masonry cavity wall consisting of solid bricks of clay, brick earth, shale, concrete or calcium silicate, with a minimum thickness of 90mm on each leaf.

Figure 60: Fire-resisting construction

1st floor

Fire-resisting floor construction to protect route above

Cavity fire barrier

Efficient smoke seal

False ceiling

Fire-resisting partition constructed up to underside of floor overhead

Protected route

Fire-resisting floor construction

Basement

There are other methods and products available which will achieve the required standard of fire resistance and may be more appropriate for the existing construction in your premises. If there is any doubt about how your building is constructed, then ask for further advice from a competent person.

Fire-resisting floors

The fire resistance of floors will depend on the existing floor construction as well as the type of ceiling finish beneath. If you need to upgrade the fire resistance of your floor it may not be desirable to apply additional fire resistance to the underside of an existing ornate ceiling. In older buildings there may be a requirement to provide fire resistance between beams and joists.

A typical example of a 30-minute fire-resisting timber floor is tongue and groove softwood of not less than 15mm finished thickness on 37mm timber joists, with a ceiling below of one layer of plasterboard to a thickness of 12.5mm with joints taped and filled and backed by supporting timber.

There are other, equally valid, methods and products available for upgrading floors. If you are in any doubt you should ask the advice of a competent person and ensure that the product is installed in accordance with instructions from the manufacturer or supplier.

Fire-resisting glazing

The most common type of fire-resisting glazing is 6mm Georgian wired glazing, which is easily identifiable. Clear fire-resisting glazing is available and can quickly be identified by a mark etched into the glass, usually in the corner of the glazed panel, to confirm its fire-resisting standard. Although this is not compulsory, the marking of glass is supported by the Glass and Glazing Federation; you should check whether the glazing will be marked accordingly before purchase. The glazing should have been installed in accordance with the manufacturer's instructions and to the appropriate standard,[66] to ensure that its fire-resisting properties are maintained.

The performance of glazed systems in terms of fire resistance and external fire exposure should, wherever possible, be confirmed by test evidence. Alternatively, where there is a lack of test information, ask for an assessment of the proposed construction from suitably qualified people.

Fire separation of voids

A common problem encountered with fire separation is fire-resisting partitions which do not extend above false ceilings to true ceiling height. This may result in unseen fire spread and a loss of vital protection to the escape routes. It is important therefore to carefully check all such partitions have been installed correctly.

CLASP and SCOLA type construction

CLASP (Consortium of Local Authorities Special Programme)* and SCOLA (Second Consortium of Local Authorities) are total or systematic methods of construction that were developed to provide consistent building quality, while reducing the need for traditional skilled labour. They consist of a metal frame upon which structural panels are fixed. This results in hidden voids through which fire may spread. It is important that cavity barriers that restrict the spread of fire are installed appropriately, especially to walls and floors that need to be fire-resisting. If you are in any doubt as to whether any remedial work will be required, then ask for advice from a competent person.

Breaching fire separation

To ensure effective protection against fire, walls and floors providing fire separation must form a complete barrier, with an equivalent level of fire resistance provided to any openings such as doors, ventilation ducts, pipe passages or refuse chutes.

The passing of services such as heating pipes or electrical cables through fire-resisting partitions leaves gaps through which fire and smoke may spread. This should be rectified by suitable fire stopping and there are many proprietary products available to suit particular types of construction. Such products should be installed by competent contractors.

Décor and surface finishes of walls, ceilings and escape routes

The materials used to line walls and ceilings can contribute significantly to the spread of flame across their surface. Most materials that are used as surface linings will fall into one of three classes of surface spread of flame. The following are common examples of acceptable materials for various situations:

Class 0: Materials suitable for circulation spaces and escape routes

- Such materials include brickwork, blockwork, concrete, ceramic tiles, plaster finishes (including rendering on wood or metal lathes), wood-wool cement slabs and mineral fibre tiles or sheets with cement or resin binding.

* www.clasp.gov.uk

Note: Additional finishes to these surfaces may be detrimental to the fire performance of the surface and if there is any doubt about this then consult the manufacturer of the finish.

Class 1: Materials suitable for use in all rooms but not on escape routes

- Such materials include all the Class 0 materials referred to above. Additionally, timber, hardboard, blockboard, particle board, heavy flock wallpapers and thermosetting plastics will be suitable if flame-retardant treated to achieve a Class 1 standard.

Class 3: Materials suitable for use in rooms of less than 30m²

- Such materials include all those referred to in Class 1, including those that have not been flame-retardant treated and certain dense timber or plywood and standard glass-reinforced polyesters.

The equivalent European classification standard will also be acceptable. Further details about internal linings and classifications are available in Approved Document B.[24] Appropriate testing procedures are detailed in BS 476-7[32] and where appropriate BS EN 13501-1.[33]

Further guidance on types of fire-resisting construction has been published by the Building Research Establishment.[34]

B2 Fire-resisting doors

Requirements of a fire-resisting door

Effective fire-resisting doors (see Figure 61) are vital to ensure that the occupants can evacuate to a place of safety. Correctly specified and well-fitted doors will hold back fire and smoke, preventing escape routes becoming unusable, as well as preventing the fire spreading from one area to another.

Fire-resisting doors are necessary in any doorway located in a fire-resisting structure. Most internal doors are constructed of timber. These will give some limited protection against fire spread, but only a purpose-built fire-resisting door that has been tested to an approved standard will provide the necessary protection. Metal fire-resisting doors are also available and specific guidance for these follows.

Figure 61: A fire-resisting and smoke stopping door

Securing device – lock, emergency exit device or panic exit device. See BS EN 12209,[87] BS EN 179[43] or BS EN 1125[42] respectively for further information.

Ideally the frame should be to the same standard as the door, purchased together as a door set.

Door closer, should conform to BS EN 1154[38] or BS EN 1155.[91]

All hinges should comply with BS EN 1935[37] annex B. Tested as part of door set to BS EN 1634-2.[39]

Vision panel should be fire-resisting glazing.

Door handles should comply with BS EN 1906[88] annex C and locks to BS EN 12209[87] annex A. Tested as part of door set to BS EN 1634-1.[35]

Intumescent strip and cold smoke seal to resist the passage of smoke and fire.

All fire-resisting doors are rated by their performance when tested to an appropriate standard. The level of protection provided by the door is measured, primarily by determining the time taken for a fire to breach the integrity (E), of the door assembly, together with its resistance to the passage of hot gases and flame.

It may be possible to upgrade the fire resistance of existing doors. Further information is available from the Building Research Establishment[76] or Timber Research and Development Association.[77]

Timber fire-resisting doors require a gap of 2-4mm between the door leaf and the frame. However larger gaps may be necessary to ensure that the door closes flush into its frame when smoke seals are fitted. Further information is available in BS 4787-1.[79] For fire-resisting purposes the gap is normally protected by installing an intumescent seal, in either the door or, preferably, the frame. The intumescent seal expands in the early stages of a fire and enhances the protection given by the door. Additional smoke seals will restrict the spread of smoke at ambient temperatures. Doors fitted with smoke seals, either incorporated in the intumescent seal or fitted separately, have their classification code suffixed with a 'S'.

The principal fire-resisting door categories are:

- E20 fire-resisting door providing 20 minutes fire resistance (or equivalent FD 20S). (Note: Many suppliers no longer provide an E20 type fire-resisting door.)

- E30 fire-resisting door providing 30 minutes fire resistance (or equivalent FD 30S).

- E60 fire-resisting door providing 60 minutes fire resistance (or equivalent FD 60S).

Timber fire-resisting doors are available that will provide up to 120 minutes fire resistance but their use is limited to more specialised conditions that are beyond the scope of this guidance.

Metal fire-resisting doors

Although the majority of fire-resisting doors are made from timber, metal fire-resisting doors, which meet the appropriate standard, can often be used for the same purpose. However, there are situations where they are more appropriate. The majority of metal fire-resisting door manufacturers will require the use of bespoke frames and hardware for their door sets.

See BS EN 1634-1[35] and BS 476-22[36] for more information.

For detailed guidance refer to Approved Document B.[24]

Glazing in fire-resisting doors

Although glazing provides additional safety in everyday use and can enhance the appearance of fire-resisting doors, it should never reduce the fire resistance of the door. The opening provided in the door for the fire-resisting glazing unit(s) and the fitting of the beading are critical, and should only be entrusted to a competent person. In nearly all cases the door and glazing should be purchased from a reputable supplier who can provide documentary evidence that the door continues to achieve the required rating.

Fire-resisting door furniture

Hinges

To ensure compliance with their rated fire performance, fire-resisting doors need to be hung with the correct number, size and quality of hinges. Normally a minimum of three hinges are needed, however the manufacturer's instructions should be closely followed. BS EN 1935[37] including Annex B, is the appropriate standard.

Alternative door mountings

Although the most common method of hanging a door is to use single axis hinges, alternative methods are employed where the door is required to be double swing or mounted on pivots for other reasons.

Floor mounted controlled door closing devices are the most common method regularly found with timber, glass and steel doors while transom mounted devices are commonly used with aluminium sections. In each case reference should be made to the fire test report for details as to compliance with the composition of the door assembly including the door mounting conditions.

Self-closing devices

All fire-resisting doors, other than those to locked cupboards and service ducts, should be fitted with an appropriately controlled self-closing device that will effectively close the door from any angle. In certain circumstances, concealed, jamb-mounted closing devices may be specified and in these cases should be capable of closing the door from any angle and against any latch fitted to the door; spring hinges are unlikely to be suitable. Further information is available in BS EN 1154.[38]

Rising butt hinges are not suitable for use as a self-closing device due to their inability to close and latch the door from any angle.

Automatic door hold-open/release devices for self-closing fire doors

These devices are designed to hold open self-closing fire doors or allow them to swing free during normal use. In the event of a fire alarm the device will then release the door automatically, allowing the self-closing mechanism to close the door.

Such devices are particularly useful in situations where self-closing doors on escape routes are used regularly by significant numbers of people, or by people with impaired mobility who may have difficulty in opening the doors.

Typical examples of such devices include:

- electro-magnetic devices fitted to the fire-resisting door which release when the fire detection and warning system operates, allowing a separate self closer to close the door;

- electro-magnetic devices within the controlled door closing device which function on the operation of the fire detection and warning system; and

- 'free swing' controlled door closing devices, which operate by allowing the door leaf to work independently of the closing device in normal conditions. An electro-magnetic device within the spring mechanism linked to the fire detection and warning system ensures that the door closes on the operation of the system.

Note: Free swing devices may not be suitable in some situations, such as corridors, where draughts are a problem and the doors are likely to swing uncontrolled, causing possible difficulty or injury to certain people e.g. those with certain disabilities, the elderly and frail, or young children.

Automatic door hold open/release devices fitted to doors protecting escape routes should only be installed in conjunction with an automatic fire detection and warning system incorporating smoke detectors, that is designed to protect the escape routes in the building (see Part 2, Section 2).

In all cases the automatic device should release the fire-resisting door allowing it to close effectively within its frame when any of the following conditions occur:

- the detection of smoke by an automatic detector;

- the actuation of the fire detection and alarm system by manual means, e.g. operation of break glass call point;

- any failure of the fire detection and alarm system; or

- any electrical power failure.

Other devices, including self-contained devices which perform a similar function, that are not connected directly to a fire alarm system and are not therefore able to meet the above criteria are available and may be acceptable where a site-specific risk assessment can show that they are appropriate. Such devices are unlikely to be suitable for use on doors protecting single stairways or other critical means of escape.

In all cases where a door hold open device is used it should be possible to close the door manually.

A site specific risk assessment should be undertaken before any type of automatic door hold open/release device is installed. If you are unsure about the suitability of such devices in your premises, you should seek the advice of a competent person.

Further guidance about automatic door hold open/release devices is given in BS EN 1155[91] or BS 5839-3.[40]

Door co-ordinators

Where pairs of doors with rebated meeting stiles are installed it is critical that the correct closing order is maintained. Door coordinators to BS EN 1158[80] should be fitted and fully operational in all cases where the doors are self-closing.

Installation and workmanship

The reliability and performance of correctly specified fire-resisting doors can be undermined by inadequate installation. It is important that installers with the necessary level of skill and knowledge are used. Accreditation schemes for installers of fire-resisting doors are available.

Fire-resisting doors and shutters will require routine maintenance, particularly to power operation and release and closing mechanisms.

Further information is available on fire-resisting doors in BS 8214.[41] If you are unsure about the quality, the effectiveness or the fitting of your fire-resisting doors consult a fire safety expert.

For further guidance on the selection and maintenance of door furniture suitable for use on timber fire resisting and escape doors refer to the Building Hardware Industry Federation (BHIF) Code of Practice – Hardware for Timber Fire and Escape Doors.[81]

B3 Door-fastening devices

The relationship between the securing of doors against unwanted entry and the ability to escape through them easily in an emergency has often proved problematical. Careful planning and the use of quality materials remain the most effective means of satisfying both of these objectives.

Any device that impedes people making good their escape, either by being unnecessarily complicated to manipulate or not being readily openable, will not be acceptable.

Guidance on fire exits starts from the position that doors on escape routes should not be fitted with any locking devices (electrically operated or otherwise). However, it is accepted that in many cases the need for security will require some form of device that prevents unlimited access, but still enables the occupants of a building or area to open the door easily if there is a fire. These devices can take many forms but, in the majority of cases, premises where there are members of the public present or others who are not familiar with the building should use panic exit bar devices (i.e. push bars or touch bars). See BS EN 1125[42] for further information.

Premises that have limited numbers of staff or others who are familiar with the building and where panic is not likely may use alternative devices (i.e. push pads or lever handles). See BS EN 179[43] for further information.

In some larger premises, when only certain staff are on the premises and there is a security issue, it may be acceptable to restrict the number of emergency exits immediately available, e.g. when only security staff are present at night, or prior to opening the premises in the morning. Staff should be made fully aware of any restrictions and the number of exits not immediately available should be limited.

Electrical locking devices

Electrically operated entry control devices have been developed for use as locking devices on fire exits. They fall into two main categories, electromechanical and electromagnetic.

- Electromechanical devices

  Electromechanical devices comprise electromechanical lock keeps and draw bolts, which can be controlled by people inside the premises by entering a code or by using 'smart cards', which have been adapted to control the exit from certain areas. These devices have been fitted in many premises and may be linked to the fire-detection and/or warning system. Experience has shown that these devices can fail to open in a number of ways. They are dependent on a spring mechanism to return the lock keep or draw bolt(s) and are liable to jam when pressure is applied to the door. It is also relatively easy to fit them incorrectly. Electromechanical locking devices are normally unacceptable on escape doors, unless they are fitted with a manual means of overriding the locking mechanism, such as a push bar, push pad or lever handle or they do not rely on a spring mechanism, fail-safe open and are not affected by pressure, in which case the criteria for electromagnetic devices should be applied.

- Electromagnetic devices

  These devices comprise a magnet and a simple fixed retaining plate with no moving parts and are therefore generally considered to be more reliable due to their inherent 'fail-safe unlocked' operation. Electromagnetic locking devices go some way to addressing the particular concerns surrounding electromechanical locking systems. The release of this type of device is controlled by the interruption of electrical current to an electromagnet either manually via a switch or other means, break-glass point (typically coloured green), or by linking to the fire-warning and/or detection system of the premises.

Time-delay devices on escape routes

A further development is the fitting of a time-delay system to the electronic door-locking device. This delays the actual opening of an exit door for a variable period following operation of the panic bar or other exit device. Periods of between five and 60 seconds can be pre-set at the manufacturing stage or can be adjusted when fitted. These are not usually acceptable for use by members of the public. However, they may be acceptable for use by staff who are familiar with their operation and are suitably trained in their use.

Management of electronic door-control devices including time delays

The use of such devices may be accepted by enforcing authorities if the responsible person can demonstrate, through a suitable risk assessment for each individual door, both the need and the adequate management controls to ensure that people can escape safely from the premises. In particular:

- Access control should not be confused with exit control. Many devices are available which control the access to the premises but retain the immediate escape facility from the premises.

- In public areas, when push bars are operated on escape doors, they should release the electromagnetic locks immediately and allow the exit doors to open.

- The requirement for exit control should be carefully assessed and should not be seen as a substitute for good management of the employees and occupants.

- All other alternatives should have been explored/evaluated prior to using these devices to ensure they do not affect the safety of occupants.

- The device should be connected to the fire warning and/or detection system.

- The device should incorporate a bypass circuit for immediate release on activation of the fire warning and/or detection system.

- Each door should be fitted with a single securing device.

- The emergency exit doors should be clearly labelled about how to operate them.

- Adequate control measures should be put in place to ensure the safety of the occupants.

The use of electronic door-locking devices should be considered with particular care in premises with a number of different occupancies. The management of a complicated system of evacuation for many different groups is unlikely to be practicable.

The technical standards in respect of sourcing, maintaining and testing must be extremely high.

When part of the management control system involves trained personnel helping others at these doors, it is vital to ensure these people are available.

The use of exit control devices should not be considered where the number of trained staff is low or where members of the public would be expected to operate the devices without help.

In premises where there may be large numbers of people, the devices should only be considered when linked to a comprehensive automatic fire-detection and warning system in accordance with BS 5839-1.[16] There should be an additional means of manually overriding the locking device at each such exit (typically a green break-glass point).

The use of time-delay systems that prevent the opening of emergency exits for a pre-set time are primarily used to improve security. These add a further layer of complexity to the fire strategy and should not be considered in public areas. They should only be used in non-public areas when all other options such as relocating valuable stock or exterior boundary management have been addressed. Their value in preventing theft is likely to be transient as the use of the manual override becomes more widely known.

British Standard 8220-3[44] gives further advice on security in buildings and while this standard does refer to electronic locking devices, it also acknowledges that the balance must remain on the side of emergency escape rather than security.

Appendix C
Historic buildings

General considerations

This appendix offers additional information about listed and historical buildings.

Fire risk assessments conducted for a place of assembly which is within a listed or historic building will need to endeavour to strike a balance between ensuring sufficient fire safety measures are in place for the safety of people, yet avoid extensive alterations and helping to maintain the character of the building.

As well as the fire risk assessment it is recommended that a general fire policy statement and manual is compiled. A person must be nominated to take responsibility for all aspects of fire safety. Usually the person charged with the management and control of the premises will be the 'responsible person' under the Order.[1]

The advice and/or consent of a building control body or any other relevant bodies (e.g. English Heritage) should form part of any fire risk assessment that impacts on the character of the building (e.g. replacement of doors, fittings, wooden panelling and decor) or material changes to existing escape routes. An ideal solution is one that is reversible, enabling the historic elements to be reinstated.

A fire safety adviser will be able to suggest alternatives to conventional fire precautions, such as:

- a fire engineering solution;

- upgrading existing doors and partitions in a sympathetic manner to improve their fire resistance; and

- considering the installation of specialist fire-detection or suppression systems.

Should the design and nature of the historic building preclude the introduction of conventional fire safety features, it will be necessary to manage the building in such a way that:

- limits the number of occupants, either staff or members of the public, inside the building;

- limits activities in the building; and

- provides adequate supervision within the building.

Liaison with the fire and rescue service

The responsible person will need to ensure effective liaison with the fire and rescue service to enable them to carry out firefighting operations. These may include information on:

- the provision of water supplies, seasonal ponds, lakes and underground tanks, and any associated pumps;

- difficult access for fire engines;

- particular hazards in the construction features of the building (including asbestos);

- the use of combustible under floor insulation;

- underground vaults ducts and voids where fire may spread unchecked;

- worn stone slabs in stairway construction; and

- the presence of cast iron columns and wrought iron beams.

Emergency planning

An important consideration for the owners and trustees is the protection of valuable artefacts and paintings from the effects of fire. However, the efficient evacuation of all occupants must take precedence over procedures for limiting damage to property and contents. Salvage work should be limited to those parts of the building not directly affected by the fire.

Fire wardens and others tasked with carrying out salvage work should have received formal training, adequate protection and be fully briefed about the health and safety risk assessment carried out to identify the dangers associated with this activity. Further detailed advice on fire safety in historic buildings can be found in the following publications:

- BS 7913: Guide to the principles of the conservation of historic buildings, British Standards Institution

- Heritage under fire: A guide to the protection of historic buildings, Fire Protection Association (for the UK Working Party on Fire Safety in Historic Buildings) 1991, ISBN 0 902167 94 4

- The Installation of Sprinkler Systems in Historic Buildings (Historic Scotland Technical Advice Note S.), Fire Protection Association (TCRE Division/Scottish Conservation Bureau, Hist.) 1998, ISBN 1 900168 63 4

- Fire Protection Measures in Scottish Historic Buildings: Advice on Measures Required to Minimise the Likelihood of Fire Starting and to Alleviate the Destructive Consequences of Fire in Historic Buildings (Technical Advice Note), TCRE Division/Scottish Conservation Bureau, Hist. 1997 ISBN 1 900168 41 3

- Fire Risk Management in Heritage Buildings (Technical Advice Note), TCRE Division/ Scottish Conservation Bureau, Hist. 2001, ISBN 1 900168 71 5

- Summary and conclusions of the report into fire protection measures for the Royal Palaces by Sir Alan Bailey following the Windsor Castle fire, 1992.

- The fire at Upton Park, The National Trust.

- Timber panelled doors and fire, English Heritage.

- Fire safety in historic town centres, English Heritage and Cheshire Fire and Rescue Service.

Appendix D

Glossary

These definitions are provided to assist the responsible person in understanding some of the technical terms used in this guide. They are not exhaustive and more precise definitions may be available in other guidance.

Term	Definition
Access room	A room through which the only escape route from an inner room passes.
Accommodation stairway	A stairway, additional to that required for means of escape purposes provided for the convenience of occupants.
Alterations notice	If your premises are considered by the enforcing authority to be high risk, they may issue an alterations notice that requires you to inform them before making any material alterations to your premises.
Alternative escape route	Escape routes sufficiently separated by either direction and space, or by fire-resisting construction to ensure that one is still available irrespective of the location of a fire.
Approved Document B (ADB)[24]	Guidance issued by Government in support of the fire safety aspects of the building regulations.
As low as reasonably practicable	Is a concept where risks should continue to be reduced until you reach a point where the cost and effort to reduce the risk further would be grossly disproportionate to the benefit achieved.
Automatic fire detection system	A means of automatically detecting the products of a fire and sending a signal to a fire warning system. See 'Fire warning'.
Basement	A storey with a floor which at some point is more than 1,200mm below the highest level of ground adjacent to the outside walls, unless, and for escape purposes only, such area has adequate, independent and separate means of escape.
Child	Anyone who is not over compulsory school age, i.e. before or just after their 16th birthday.
Class 0, 1 or 3 surface spread of flame	Classes of surface spread of flame for materials needed to line the walls and ceilings of escape routes. See Appendix B for further information.
Combustible material	A substance that can be burned.
Compartment wall and/or floor	A fire-resisting wall or floor that separates one fire compartment from another.
Competent person	A person with enough training and experience or knowledge and other qualities to enable them properly to assist in undertaking the preventive and protective measures.

Term	Definition
Dangerous substance	1. A substance which because of its physico-chemical or chemical properties and the way it is used or is present at the workplace creates a risk. 2. A substance subject to the Dangerous Substances and Explosive Atmosphere Regulations 2002 (DSEAR).
Dead end	Area from which escape is possible in one direction only.
Direct distance	The shortest distance from any point within the floor area to the nearest storey exit, or fire-resisting route, ignoring walls, partitions and fixings.
Domestic premises	Premises occupied as a private dwelling, excluding those areas used in common by the occupants of more than one such dwelling.
Emergency escape lighting	Lighting provided to illuminate escape routes that will function if the normal lighting fails.
Enforcing authority	The fire and rescue authority or any other authority specified in Article 25 of the Regulatory Reform (Fire Safety) Order 2005.[1]
Escape route	Route forming that part of the means of escape from any point in the premises to a final exit.
Evacuation lift	A lift that may be used for the evacuation of people with disabilities, or others, in a fire.
External escape stair	Stair providing an escape route, external to the building.
Fail-safe	Locking an output device with the application of power and having the device unlock when the power is removed. Also known as fail unlock, reverse action or power locked.
False alarm	A fire signal, usually from a fire warning system, resulting from a cause other than fire.
Final exit	An exit from a building where people can continue to disperse in safety and where they are no longer at danger from fire and/or smoke.
Fire compartment	A building, or part of a building, constructed to prevent the spread of fire to or from another part of the same building or an adjoining building.
Fire door	A door or shutter, together with its frame and furniture, provided for the passage of people, air or goods which, when closed is intended to restrict the passage of fire and/or smoke to a predictable level of performance.
Firefighting lift	A lift, designed to have additional protection, with controls that enable it to be used under the direct control of the fire and rescue service when fighting a fire.
Firefighting shaft	A fire-resisting enclosure containing a firefighting stair, fire mains, firefighting lobbies and if provided, a firefighting lift.

Term	Definition
Firefighting stairway	See firefighting shaft.
Fire resistance	The ability of a component or construction of a building to satisfy, for a stated period of time, some or all of the appropriate criteria of relevant standards. (Generally described as 30 minutes fire-resisting or 60 minutes fire-resisting.) See BS EN 1363-1,[45] BS 476-7[32] and associated standards for further information.
Fire safety manager	A nominated person with responsibility for carrying out day-to-day management of fire safety. (This may or may not be the same as the 'responsible person'.)
Fire safety strategy	A number of planned and co-ordinated arrangements designed to reduce the risk of fire and to ensure the safety of people if there is a fire.
Fire stopping	A seal provided to close an imperfection of fit or design tolerance between elements or components, to restrict the passage of fire and smoke.
Fire-warning system	A means of alerting people to the existence of a fire. (See automatic fire detection system.)
Flammable material	Easily ignited and capable of burning rapidly.
Highly flammable	Generally liquids with a flashpoint of below 21°C. (The Chemicals Hazard Information and Packaging for Supply Regulations 2002[47] (CHIP) give more detailed guidance.)
Hazardous substance	1. See Dangerous substance. 2. A substance subject to the Control of Substances Hazardous to Health Regulations 2002 (COSHH).
Inner room	A room from which escape is possible only by passing through another room (the access room).
Licensed premises	Any premises that require a licence under any statute to undertake trade or conduct business activities.
Material change	An alteration to the premises, process or service which significantly affects the level of risk to people from fire in those premises.
Means of escape	Route(s) provided to ensure safe egress from the premises or other locations to a place of total safety.
Phased evacuation	A system of evacuation in which different parts of the premises are evacuated in a controlled sequence of phases, those parts of the premises expected to be at greatest risk being evacuated first.
Place of reasonable safety	A place within a building or structure where, for a limited period of time, people will have some protection from the effects of fire and smoke. This place, usually a corridor or stairway, will normally have a minimum of 30 minutes fire resistance and allow people to continue their escape to a place of total safety.

Term	Definition
Place of total safety	A place, away from the premises, in which people are at no immediate danger from the effects of a fire.
Premises	Any place, such as a building and the immediate land bounded by any enclosure of it, any tent, moveable or temporary structure or any installation or workplace.
Protected lobby	A fire-resisting enclosure providing access to an escape stairway via two sets of fire doors and into which no room opens other than toilets and lifts.
Protected stairway	A stairway which is adequately protected from the rest of the building by fire-resisting construction.
Protected route	An escape route which is adequately protected from the rest of the building by fire-resisting construction.
Refuge	A place of reasonable safety in which a disabled person and others who may need assistance may rest or wait for assistance before reaching a place of total safety. It should lead directly to a fire-resisting escape route.
Responsible person	The person ultimately responsible for fire safety as defined in the Regulatory Reform (Fire Safety) Order 2005.[1]
Relevant persons	Any person lawfully on the premises and any person in the immediate vicinity, but does not include firefighters carrying out firefighting duties.
Self-closing device	A device that is capable of closing the door from any angle and against any latch fitted to the door.
Significant finding	A feature of the premises, from which the fire hazards and persons at risk are identified. The actions you have taken or will take to remove or reduce the chance of a fire occuring or the spread of fire and smoke. The actions people need to take in case of fire. The necessary information, instruction and training needed and how it will be given.
Smoke alarm	Device containing within one housing all the components, except possibly the energy source, for detecting smoke and giving an audible alarm.
Staged fire alarms	A fire warning which can be given in two or more stages for different purposes within a given area (i.e. notifying staff, stand by to evacuate, full evacuation).
Storey exit	A final exit or a doorway giving direct access into a protected stairway, firefighting lobby, or external escape route.
Travel distance	The actual distance to be travelled by a person from any point within the floor area to the nearest storey exit or final exit, having regard to the layout of walls, partitions and fixings.

Term	Definition
Vision panel	A transparent panel in a wall or door of an inner room enabling the occupant to become aware of a fire in the access area during the early stages.
Way guidance	low mounted luminous tracks positioned on escape routes in combination with exit indicators, exit marking and intermediate direction indicators along the route, provided for use when the supply to the normal lighting fails, which do not rely on an electrical supply for their luminous output.
Where necessary	The Order requires that fire precautions (such as firefighting equipment, fire detection and warning, and emergency routes and exits) should be provided (and maintained) 'where necessary'. What this means is that the fire precautions you must provide (and maintain) are those which are needed to reasonably protect relevant persons from risks to them in case of fire. This will be determined by the findings of your risk assessment including the preventative measures you have or will have taken. In practice, it is very unlikely, that a properly conducted fire risk assessment, which takes into account all the matters relevant for the safety of persons in case of fire, will conclude that no fire precautions (including maintenance) are necessary.
Young person	(a) A person aged 16 years, from the date on which he attains that age until and including the 31st August which next follows that date. (b) A person aged 16 years and over who is undertaking a course of full-time education at a school or college which is not advanced education. (c) A person aged 16 years and over who is undertaking approved training that is not provided through a contract of employment. For the purposes of paragraphs (b) and (c) the person: (a) shall have commenced the course of full-time education or approved training before attaining the age of 19 years; and (b) shall not have attained the age of 20 years.

References

The following documents are referenced in this guide. Where dated, only this version applies. Where undated, the latest version of the document applies.

1   Regulatory Reform (Fire Safety) Order 2005, SI 2005/1541. The Stationery Office, 2005. ISBN 0 11 072945 5.

2   Fire Precautions Act 1971 (c 40). The Stationery Office, 1971. ISBN 0 10 544071 X.

3   Fire Precautions (Workplace) Regulations 1997. SI 1997/1840. The Stationery Office, 1997. ISBN 0 11 064738 6.

4   Fire Precautions (Workplace) (Amendment) Regulations 1999. SI 1999/1877. The Stationery Office, 1999. ISBN 0 11 082882 8.

5   Health and Safety (Safety Signs and Signals) Regulations 1996. SI 1996/341. The Stationery Office, 1996. ISBN 0 11 054093 X.

6   *Safety signs and signals. The Health and Safety (Safety Signs and Signals) Regulations 1996. Guidance on regulations,* L64. HSE Books, 1996. ISBN 0 7176 0870 0.

7   Dangerous Substances and Explosive Atmospheres Regulations 2002, SI 2002/2776. The Stationery Office, 2002. ISBN 0 11 042957 5.

8   *Dangerous substances and explosive atmospheres. Dangerous Substances and Explosive Atmospheres Regulations 2002. Approved code of practice and guidance,* L138. HSE Books, 2003. ISBN 0 7176 2203 7.

9   *Storage of full and empty LPG cylinders and cartridges. Code of Practice 7.* LP Gas Association, 2000. Available from LP Gas Association, Pavilion 16, Headlands Business Park, Salisbury Road, Ringwood, Hampshire BH24 3PB.

10 *Maintaining portable electrical equipment in offices and other low-risk environments,* INDG236. HSE Books, 1996. (ISBN 0 7176 1272 4 single copy free or priced packs of 10).

11 Construction (Health, Safety and Welfare) Regulations 1996, SI 1996/1592. The Stationery Office, 1996. ISBN 0 11 035904 6.

12 *A guide to the Construction (Health, Safety and Welfare) Regulations 1996,* INDG220. HSE Books, 1996. (ISBN 0 7176 1161 2 single copy free or priced packs of 10).

    Health and safety in construction, HSG150 (second edition). HSE Books, 2001. ISBN 0 7176 2106 5.

13 Disability Discrimination Act 1995 (c 50). The Stationery Office, 1995. ISBN 0 10 545095 2.

14 BS 8300: *The design of buildings and their approaches to meet the needs of disabled people. Code of practice.* British Standards Institution. ISBN 0 580 38438 1.

15 ODPM/CACFOA/BFPSA guidance on reducing false alarms.

16 BS 5839-1: *Fire detection and alarm systems for buildings. Code of practice for system design, installation, commissioning and maintenance.* British Standards Institution. ISBN 0 580 40376 9.

17 Manual Handling Operations Regulations 1992, SI 1992/2793. The Stationery Office, 1992. ISBN 0 11 025920 3.

18 BS 5306-8: *Fire extinguishing installations and equipment on premises. Selection and installation of portable fire extinguishers. Code of practice* British Standards Institution ISBN 0 580 33203 9

19 BS 5306-3: *Fire extinguishing installations and equipment on premises. Code of practice for the inspection and maintenance of portable fire extinguishers.* British Standards Institution. ISBN 0 5808 42865 6.

20 BS 7863: *Recommendations for colour coding to indicate the extinguishing media contained in portable fire extinguishers.* British Standards Institution. ISBN 0 580 25845 9.

21 BS EN 671-3: *Fixed firefighting systems. Hose systems. Maintenance of hose reels with semi-rigid hose and hose systems with lay-flat hose.* British Standards Institution. ISBN 0 580 34112 7.

22 BS EN 12845: *Fixed firefighting systems. Automatic sprinkler systems. Design, installation and maintenance.* British Standards Institution. ISBN 0 580 44770 7.

23 Workplace (Health, Safety and Welfare) Regulations 1992, SI 1992/3004. The Stationery Office, 1992. ISBN 0 11 025804 5.

24 *The Building Regulations 2000: Approved Document B fire safety*. The Stationery Office. ISBN 0 11 753911 2.

25 Local Government (Miscellaneous Provisions) Act 1982 (c 30). The Stationery Office, 1982. ISBN 0 10 543082 X.

26 BS 5395-2: *Stairs, ladders and walkways. Code of practice for the design of industrial type stairs, permanent ladders and walkways*. British Standards Institution. ISBN 0 580 14706 1.

27 BS 5266-8: *Emergency lighting. Code of practice for emergency escape lighting systems*. British Standards Institution.

27A BS 5266-1: *Emergency lighting. Code of practice for the emergency lighting of premises*. British Standards Institution.

28 BS EN 1838: *Lighting applications. Emergency lighting*. British Standards Institution. ISBN 0 580 32992 5.

29 BS 5499-5: *Graphical symbols and signs. Safety signs, including fire safety signs. Signs with specific safety meanings* British Standards Institution

29A BS 5499-4: *Safety signs, including fire safety signs. Code of practice for escape route signing*. British Standards Institution.

30 BS 7974: *Application of fire safety engineering principles to the design of buildings. Code of practice*. British Standards Institution. ISBN 0 580 38447 0.

31 *Successful health and safety management*, HSG65 (second edition). HSE Books, 1997. ISBN 0 7176 1276 7.

32 BS 476-7: *Fire tests on building materials and structures. Method of test to determine the classification of the surface spread of flame of products*. British Standards Institution.

33 BS EN 13501-1: *Fire classification of construction products and building elements. Classification using test data from reaction to fire tests*. British Standards Institution.

34 *Guidelines for the construction of fire-resisting structural elements*, BRE 128. Building Research Establishment, 1988.

35 BS EN 1634-1: *Fire resistance tests for door and shutter assemblies. Fire doors and shutters*. British Standards Institution. ISBN 0 580 32429 X.

36 BS 476-22: *Fire tests on building materials and structures. Methods for determination of the fire resistance of non-loadbearing elements of construction*. British Standards Institution. ISBN 0 580 15872 1.

37 BS EN 1935: *Building hardware. Single-axis hinges. Requirements and test methods*. British Standards Institution. ISBN 0 580 39272 4.

38 BS EN 1154: *Building hardware. Controlled door closing devices. Requirements and test methods*. British Standards Institution. ISBN 0 580 27476 4.

39 BS EN 1634-2: *Fire resistance tests for door and shutter assemblies. Part 2. Fire door hardware. Building hardware for fire-resisting doorsets and openable windows*. British Standards Institution.

40 BS 5839-3: *Fire detection and alarm systems for buildings. Specification for automatic release mechanisms for certain fire protection equipment*. British Standards Institution. ISBN 0 580 15787 3.

41 BS 8214: *Code of practice for fire door assemblies with non-metallic leaves*. British Standards Institution. ISBN 0 580 18871 6.

42 BS EN 1125: *Building hardware. Panic exit devices operated by a horizontal bar. Requirements and test methods*. British Standards Institution.

43 BS EN 179: *Building hardware. Emergency exit devices operated by a lever handle or push pad. Requirements and test methods*. British Standards Institution. ISBN 0 580 28863 3.

44 BS 8220: *Guide for security of buildings against crime*. British Standards Institution. ISBN 0 580 23692 7.

45 BS EN 1363-1: *Fire resistance tests. General requirements*. British Standards Institution.

46 Chemicals (Hazard Information and Packaging for Supply) Regulations 2002, SI 2002/1689. The Stationery Office, 2002. ISBN 0 11 042419 0.

47 Licensing Act 2003. The Stationery Office. ISBN 0 10 541703 3.

48 *Guide to safety at sports grounds* (fourth edition). Department of National Heritage/ The Scottish Office, 1997.

49 *Guide to health, safety and welfare at pop concerts and similar events.* Health and Safety Commission/Home Office/The Scottish Office, 1993.

50 BS 5588-10: *Fire precautions in the design, construction and use of buildings. Code of practice for shopping complexes.* British Standards Institution.

51 BS 5588-12: *Fire precautions in the design, construction and use of buildings – Part 12: Managing fire safety.* British Standards Institution.

52 *Code of practice for pyrotechnics and smoke effects.* Association of British Theatre Technicians.

53 *Smoke and vapour effects used in entertainment: Entertainment Information Sheet No. 3.* Health and Safety Executive.

54 Safety at Sports Grounds Act 1975. HMSO, 1975.

55 The Electricity at Work Regulations 1989. SI 1989/635.

56 The Electrical Equipment (Safety) Regulations 1994. SI 1994/3260.

57 The Construction (Design and Management) Regulations 1994 (CONDAM/CDM Regs). HMSO, 1994.

58 *Construction Information Sheet No. 51: Construction fire safety.* Health and Safety Executive.

59 *Fire safety in construction work,* HSG 168. Health and Safety Executive. ISBN 0 7176 1332 1.

60 *Fire prevention on construction sites: The joint code of practice on the protection from fire of construction sites and buildings undergoing renovation* (fifth edition). Fire Protection Association and Construction Federation, 2000. ISBN 0 902167 39 1.

61 BS 7157: *Method of test for ignitability of fabrics used in the construction of large tented structures.* British Standards Institution.

62 BS 6661: *Guide for the design, construction and maintenance of single-skin air supported structures.* British Standards Institution.

63 *Design, construction, specification and fire management of insulated envelopes for temperature controlled environments.* International Association for Cold Storage Construction.

64 The Furniture and Furnishings (Fire) (Safety) Regulations 1988. The Stationery Office. ISBN 0 11 087324 6.

65 The Furniture and Furnishings (Fire) (Safety) (Amendment) Regulations 1989. The Stationery Office. ISBN 0 11 098358 0.

66 BS 5867-2: *Specification for fabrics for curtains and drapes. Flammability requirements.* British Standards Institution.

67 BS 1892-2: *Gymnasium equipment. Particular requirements. Specification for boxing rings.* British Standards Institution.

68 BS 5588-6: *Fire precautions in the design, construction and use of buildings. Code of practice for places of assembly.* British Standards Institution.

69 BS 5306-2: *Fire extinguishing installations and equipment on premises. Specification for sprinkler systems.* British Standards Institution.

70 BS 5588-5: *Fire precautions in the design, construction and use of buildings. Access and facilities for firefighting.* British Standards Institution.

71 *The Building Regulations 1991: Approved Document M access to and use of buildings* (2004 edition).

72 BS 5588-8: *Fire precautions in the design, construction and use of buildings. Code of practice for means of escape for disabled people.* British Standards Institution.

73 *CIBSE Guide Volume E: Fire engineering.* Chartered Institution of Building Services Engineers, 1997.

74 *Design methodologies for smoke and heat exhaust ventilation,* Report 368. BRE, 1999.

75 *A guide to best practice in the specification and use of fire-resistant glazed systems.* Glass and Glazing Federation, 2005.

76 *Increasing the fire resistance of existing timber doors,* Information Paper 8/82. BRE.

77 *Fire-resisting doorsets by upgrading,* Wood Information Sheet 1-32. Timber Research and Development Association.

78 BS 5266-6: *Emergency lighting. Code of practice for non-electrical low mounted way guidance systems for emergency use. Photoluminescent systems.* British Standards Institution.

79 BS 4787-1: *Internal and external wood doorsets, door leaves and frames. Specification for dimensional requirements.* British Standards Institution.

80 BS EN 1158: *Building hardware. Door coordinator devices. Requirement and test methods.* British Standards Institution

81 *Code of practice: Hardware for timber fire and escape doors.* The British Hardware Industry Federation (BHIF).

82 BS 5852: *Methods of test for the assessment of the ignitability of upholstered seating by smouldering and flaming ignition.* British Standards Institution.

83 BS 7176: *Specification for resistance to ignition of upholstered furniture for non-domestic seating by testing composites.* British Standards Insitution.

84 BS EN 1101: *Textiles and textile products. Burning behaviour. Curtains and drapes. Detailed procedure to determine the ignitability of vertically orientated speciments (small flame).* British Standards Institution.

85 BS EN 1102: *Textiles and textile products. Burning behaviour. Curtains and drapes. Detailed procedure to determine the flame spread of vertically orientated specimens.* British Standards Institution.

86 BS EN 3-7: *Portable fire extinguishers. Characteristics, performance requirements and test methods.* British Standards Institution.

87 BS EN 12209: *Building hardware. Locks and latches. Mechanically operated locks, latches and locking plate.* British Standards Institution.

88 BS EN 1906: *Building hardware. Lever handles and knob furniture. Requirements and test methods.* British Standards Institution.

89 *LPC rules for automatic sprinkler installations.* The Fire Protection Association, 2003.

90 BS 5287: *Specification for assessment and labelling of textile floor covering tested to BS 4790.* British Standards Institution.

91 BS EN 1155: *Building hardware. Electrically powered hold open devices for swing doors. Requirements and test methods.* British Standards Institution.

Further reading

The latest versions of all documents listed in this section should be used, including any amendments.

Any views expressed in these documents are not necessarily those of the DCLG.

BS 4422	Fire. Vocabulary. British Standards Institution.
BS PD 6512-3	Use of elements of structural fire protection with particular reference to the recommendations given in BS 5588 *Fire precautions in the design and construction of buildings. Guide to the fire performance of glass.* British Standards Institution.
BS EN 81	Safety rules for the construction and installation of lifts. British Standards Institution.
BS EN 81-70	Safety rules for the construction and installation of lifts. Particular applications for passenger and goods passenger lifts. Accessibility to lifts for persons including persons with disability. British Standards Institution.
BS 5041-1	Fire hydrant systems equipment. Specification for landing valves for wet risers. British Standards Institution.
BS 5041-2	Fire hydrant systems equipment. Specification for landing valves for dry risers. British Standards Institution.
BS 5041-3	Fire hydrant systems equipment. Specification for inlet breechings for dry riser inlets. British Standards Institution.
BS 5041-4	Fire hydrant systems equipment. Specification for boxes for landing valves for dry risers. British Standards Institution.
BS 5041-5	Fire hydrant systems equipment. Specification for boxes for foam inlets and dry riser inlets. British Standards Institution.
BS 9990	Code of practice for non-automatic firefighting systems in buildings. British Standards Institution.
BS 7944	Type 1 heavy duty fire blankets and type 2 heavy duty heat protective blankets. British Standards Institution.
BS EN 1869	Fire blankets. British Standards Institution.
BS ISO 14520-1	Gaseous fire-extinguishing systems. Physical properties and system design. General requirements. British Standards Institution.
BS EN 60598-1	Luminaires. General requirements and tests. British Standards Institution.
BS 5266-2	Emergency lighting. Code of practice for electrical low mounted way guidance for emergency use. British Standards Institution.

BS 5499-1	Graphical symbols and signs. Safety signs, including fire safety signs. Specification for geometric shapes, colours and layout. British Standards Institution.
BS EN 1634-3	Fire resistance tests for door and shutter assemblies. Smoke control doors and shutters. British Standards Institution.
Draft BS EN 14637	Building hardware. Electrically controlled hold-open systems for fire/smoke door assemblies. Requirements, test methods, application and maintenance. (Consultation document). British Standards Institution.
BS EN 45020	Standardization and related activities. General vocabulary. British Standards Institution.
ISO 13784-2	Reaction to fire tests for sandwich panel building systems Part 2: Test method for large rooms. British Standards Institution.
BS 5268-4.2	Structural use of timber. Fire resistance of timber structures. Recommendations for calculating fire resistance of timber stud walls and joisted floor constructions. British Standards Institution.

Technical standards for places of entertainment. District Surveyors Association/Association of British Theatre Technicians.

Management of Health and Safety at Work Regulations 1999. SI 1999/3242.

Design principles of fire safety. The Stationery Office, 1996. ISBN 0 11 753045 X.

Chemicals (Hazard Information and Packaging for Supply) Regulations 2002, SI 2002/1689. The Stationery Office, 2002. ISBN 0 11 042419 0. Supporting guides: *The idiot's guide to CHIP 3: Chemicals (Hazard Information and Packaging for Supply) Regulations 2002*, INDG350. HSE Books, 2002. ISBN 0 7176 2333 5 (single copy free or priced packs of 5); *CHIP for everyone,* HSG228. HSE Books, 2002. ISBN 0 7176 2370 X.

Guidance on the acceptance of electronic locks to doors required for means of escape. The Chief and Assistant Chief Fire Officers' Association.

Ensuring best practice for passive fire protection in buildings. Building Research Establishment, 2003. ISBN 1 870409 19 1.

Smoke shafts protecting fire shafts: their performance and design, BRE Project Report 79204. Building Research Establishment, 2002.

Fire safety of PTFE-based material used in building, BRE Report 274. Building Research Establishment, 1994. ISBN 0 851256 53 8.

Fires and human behaviour. David Fulton Publishers, 2000. ISBN 1 85346 105 9.

Management of health and safety at work. Management of Health and Safety at Work Regulations 1999. Approved code of practice and guidance, L21 (second edition). HSE Books, 2000. ISBN 0 7176 2488 9.

Index

Page numbers in italics refer to information in Figures or Tables.